Memoirs of the American Mathematical Society
Number 312

Josefina Casasayas and Jaume Llibre

Qualitative analysis of the anisotropic Kepler problem

Published by the
AMERICAN MATHEMATICAL SOCIETY
Providence, Rhode Island, USA

November 1984 · Volume 52 · Number 312 (second of 3 numbers)

MEMOIRS of the American Mathematical Society

This journal is designed particularly for long research papers (and groups of cognate papers) in pure and applied mathematics. It includes, in general, longer papers than those in the TRANSACTIONS.

Mathematical papers intended for publication in the Memoirs should be addressed to one of the editors. Subjects, and the editors associated with them, follow:

Ordinary differential equations, partial differential equations and applied mathematics to JOEL A. SMOLLER, Department of Mathematics, University of Michigan, Ann Arbor, MI 48109.

Complex and harmonic analysis to LINDA PREISS ROTHSCHILD, Department of Mathematics, University of California at San Diego, LaJolla, CA 92093

Abstract analysis to WILLIAM B. JOHNSON, Department of Mathematics, Texas A&M University, College Station, TX 77843-3368

Algebra, algebraic geometry and number theory to LANCE W. SMALL, Department of Mathematics, University of California at San Diego, LaJolla, CA 92093

Logic, set theory and general topology to KENNETH KUNEN, Department of Mathematics, University of Wisconsin, Madison, WI 53706

Topology to WALTER D. NEUMANN, Mathematical Sciences Research Institute, 2223 Fulton Street, Berkeley, CA 94720

Global analysis and differential geometry to TILLA KLOTZ MILNOR, Department of Mathematics, Hill Center, Rutgers University, New Brunswick, NJ 08903

Probability and statistics to DONALD L. BURKHOLDER, Department of Mathematics, University of Illinois, Urbana, IL 61801

Classical analysis to PETER W. JONES, Department of Mathematics, University of Chicago, Chicago, IL 60637

Combinatorics and number theory to RONALD GRAHAM, Mathematical Sciences Research Center, AT&T Bell Laboratories, 600 Mountain Avenue, Murray Hill, NJ 07974

All other communications to the editors should be addressed to the Managing Editor, R. O. WELLS, JR., Department of Mathematics, Rice University, Houston, TX 77251

MEMOIRS are printed by photo-offset from camera-ready copy fully prepared by the authors. Prospective authors are encouraged to request booklet giving detailed instructions regarding reproduction copy. Write to Editorial Office, American Mathematical Society, P. O. Box 6248, Providence, Rhode Island 02940. For general instructions, see last page of Memoir.

SUBSCRIPTION INFORMATION. The 1984 subscription begins with Number 289 and consists of six mailings, each containing one or more numbers. Subscription prices for 1984 are $148 list; $74 member. A late charge of 10% of the subscription price will be imposed upon orders received from nonmembers after January 1 of the subscription year. Subscribers outside the United States and India must pay a postage surcharge of $10; subscribers in India must pay a postage surcharge of $15. Each number may be ordered separately; *please specify number* when ordering an individual number. For prices and titles of recently released numbers, refer to the New Publications sections of the NOTICES of the American Mathematical Society.

BACK NUMBER INFORMATION. For back issues see the AMS Catalogue of Publications.

TRANSACTIONS of the American Mathematical Society

This journal consists of shorter tracts which are of the same general character as the papers published in the MEMOIRS. The editorial committee is identical with that for the MEMOIRS so that papers intended for publication in this series should be addressed to one of the editors listed above.

Subscriptions and orders for publications of the American Mathematical Society should be addressed to American Mathematical Society, P. O. Box 1571, Annex Station, Providence, R. I. 02901. *All orders must be accompanied by payment.* Other correspondence should be addressed to P. O. Box 6248, Providence, R. I. 02940.

MEMOIRS of the American Mathematical Society (ISSN 0065-9266) is published bimonthly (each volume consisting usually of more than one number) by the American Mathematical Society at 201 Charles Street, Providence, Rhode Island 02904. Second Class postage paid at Providence, Rhode Island 02940. Postmaster: Send address changes to Memoirs of the American Mathematical Society, American Mathematical Society, P. O. Box 6248, Providence, RI 02940.

TABLE OF CONTENTS

ABSTRACT. We give a qualitative analysis of the flow of the anisotropic Kepler problem described by the Hamiltonian system:

$$\dot{q} = M^{-1}p \ , \qquad \dot{p} = -q/\|q\|^3 \qquad\qquad (1)$$

where $(q,p) \ \varepsilon \ (R^2/\{0\}) \times R^2$, $M^{-1} = \begin{pmatrix} \mu & 0 \\ 0 & 1 \end{pmatrix}$ is the mass matrix and the parameter μ belongs to $[1,\infty)$. It was introduced by Gutzwiller and later it was studied by Devaney. When $\mu = 1$ it is the Kepler problem (an integrable system) and we show the global orbit structure by taking into account the blow up of the singularities at $q = 0$ and $\|q\| = \infty$. When $\mu \ \varepsilon \ (9/8,\infty)$ symbolic dynamic allows us to classify the solutions of (1). In fact, we prove that the dynamic behaviour contains a subshift with an infinite alphabet. The symbols of this alphabet takes into account the symmetries. For each periodic sequence of this subshift we show the existence of a symmetric periodic orbit which realizes it. The transition from $\mu = 1$ (integrable) to $\mu > 9/8$ (chaotic) is such that the chaos does not appear until $\mu = 9/8$.

AMS Subject classfication (1980): 34C35, 58F15

Library of Congress Cataloging in Publication Data

Casasayas, Josefina, 1957-
    Qualitative analysis of the anisotropic Kepler problem.

    (Memoirs of the American Mathematical Society,
ISSN 0065-9266 ; no. 312 (Nov. 184))
    "November 1984, volume 52, number 312 (second of three numbers)."
    Bibliography: p.
    1. Differential equations. 2. Differentiable dynamical systems. 3. Hamiltonian systems. I. Llibre, Jaume. II. Title. III. Title: Anisotropic Kepler problem. IV. Series: Memoirs of the American Mathematical Society ; no. 312.
    QA3.A57 no. 312 [QA372] 510 s [515.3'5] 84-18521
    ISBN 0-8218-2309-4

0 INTRODUCTION

The anisotropic Kepler problem was introduced by Gutzwiller as a classical mechanical system which approximates the following quantum mechanical system: the study of bound states of an electron near a donor impurity of a semiconductor. For more details on the physical connections we refer to [G1,2,3,4,5] .

As it is known the anisotropic Kepler problem exhibits many qualitative phenomena of interest in the theory of differential equations such as non- integrability and chaotic behaviour, see [G5,6] and [D2,3] . This paper is essentially devoted to the qualitative analysis of this problem, and also surveys the recent techniques and results from it.

The anisotropic Kepler problem is a one parameter family (of parameter μ) of Hamiltonian systems with two degrees of freedom. The configuration space for the system is $Q = \mathbb{R}^2 \setminus \{0\}$ with coordinates $q = (q_1, q_2)$, and the phase space is the tangent bundle to Q which we denote by $TQ = (\mathbb{R}^2 \setminus \{0\}) \times \mathbb{R}^2$. We use coordinates $p = (p_1, p_2)$ in each fiber. Then the Hamiltonian is,

$$H(q,p) = (p^t M^{-1} p)/2 + V(q),$$

where H is defined on TQ, the mass matrix $M^{-1} = \begin{pmatrix} \mu & 0 \\ 0 & 1 \end{pmatrix}$, and the potential energy $V(q) = -1/\|q\|$. The associated Hamilton equations are,

$$\dot{q} = M^{-1} p , \qquad\qquad (1)$$
$$\dot{p} = -q/\|q\|^3 .$$

Of course, the Hamiltonian H is an integral of (1). So, orbits of (1) lie on the energy levels of H. In (II.1) we note that it is sufficient to study the cases H=-1, H=0, and H=1.

When $\mu=1$, (1) becomes the Kepler problem, which is an integrable system. It is known that when $\mu > 1$ system (1) does not have any real analytic integral independent on the energy (see [D2] and [Mo]).

Note that for $\mu > 1$ the q_2-axis is a "heavy" axis, this means that the orbits oscillate more and more rapidly about the q_2-axis as μ increases.

For every energy level system (1) has a singularity at $q=0$. It has been studied by Devaney in [D2,5] by using the blow up techniques of McGehee [Mc] . For non-negative energy levels we have another singularity at $\|q\| = \infty$; again, blow up techniques can be applied, see Lacomba-Simó in [LS] .

The blow up method replaces the singularity by an invariant boundary manifold and the system extends over it. So, the knowledge of the flow on this boundary allows to study the behaviour of the orbits near the singularity. Thus, the invariant boundaries glued to $q=0$ and $\|q\|=\infty$ are called the collision manifold and the infinity manifold, respectively.

In system (1) the blow up of the singularities is essential in order to make the qualitative analysis of the flow. Thus, in Chapter I we describe the global behaviour of the orbit structure of the Kepler problem by taking into account the blow up of the singularities.

The first part of Chapter II is also devoted to the singularities of the anisotropic Kepler problem. In the remaining part we analize the homothetic orbits. Since these orbits are heteroclinic and transversal, they play a major part in the qualitative analysis. Transversality was proved by Devaney for negative energy levels [D4] ; we extend it to non-negative energy levels in (II.6). In (II.7) we give the global behaviour of the flow on the collision manifold for all $\mu > 1$. This improves the results of Devaney in [D2] .

As it was observed in [LS] the global orbit structure in the zero energy level can be obtained from the global flow on the collision manifold. This is shown in (III.1). The asymptotic behaviour of the orbits in the positive energy levels is given in (III.2).

In the non-negative energy levels we do not have recurrent orbits. So, the interesting case is H<0. In order to describe recurrent motions it is useful to introduce symbolic dynamics.

Gutzwiller and Devaney use symbolic dynamics to classify the possible types of orbits in the anisotropic Kepler problem, see [D5,pp.292-297] . As they said, their symbols do not take into account the symmetries of the problem. In this paper symbolic dynamics includes the symmetries, see Theorems IV.17 and IV.17' given in (IV.7).

Proofs of these theorems need the qualitative analysis of the intersection of the stable and unstable invariant manifolds of the equilibrium points of the problem with the surface of section $d/dt(\|q\|)=0$. Such an analysis is the key point of this study and it is made in the first five sections of Chapter IV.

In fact, theorems of Gutzwiller, Devaney, IV.17 and IV.17' prove the existence of a subshift with an infinite alphabet as a "subsystem" of an adequate Poincaré map for $\mu>9/8$.

In Chapter V we describe the transition from the integrable case $\mu=1$ to the chaotic one $\mu>9/8$. That is, (V.2) shows that the chaotic behaviour observed for $\mu>9/8$ is completely lost when $1\leq\mu\leq9/8$.

In Chapter VI we study the symmetric periodic orbits with respect to the six symmetries of the problem. For the simplest ones we describe their geometry, see Theorem VI.5. Also, for each periodic sequence of the subshift given in Theorem IV.17 and IV.17', we show the existence of a symmetric periodic orbit which realizes it, see Theorem VI.6.

This paper has allowed to J.Casasayas to obtain her Degree of Doctor of Philosophy (Mathematics) for the University of Barcelona.

The authors wish to acknowledge the many helpful discussions with Dr. C.Simó.

I. THE KEPLER PROBLEM

(I.1) Formulation

We consider the Hamiltonian,

$$H(q,p) = \frac{1}{2} p^t M^{-1} p + V(q)$$

where H is defined on $(\mathbb{R}^2 - \{0\}) \times \mathbb{R}^2$, M is the identity 2x2 matrix, and the potential energy $V(q) = -\|q\|^{-1} = -(q_1^2 + q_2^2)^{-\frac{1}{2}}$. Then the Hamiltonian equations become :

$$\begin{aligned}
\dot{q} &= p \\
\dot{p} &= -\text{grad } V(q)
\end{aligned} \tag{1}$$

where the dot denotes the derivative with respect to t.

These differential equations define the Kepler problem.

Hamiltonian equations (1) are integrable; that is, there exists another integral, the <u>angular momentum</u>, $C(q,p) = q \wedge p$, in involution with H (see [AM]).

We note that equations (1) have two singularities, one when q=0 and the other when $\|q\| = \infty$.

A solution $(q(t), p(t))$ of (1) is called a <u>collision</u> (resp. <u>ejection</u>) <u>solution</u> if there exists t_0 such that $q(t) \to 0$ when $t \downarrow t_0$ (resp. $t \uparrow t_0$). We say that a solution $(q(t), p(t))$ is an <u>escape</u> (resp. <u>capture</u>) <u>solution</u> if $\|q(t)\| \to \infty$ when $t \to \infty$ (resp. $t \to -\infty$).

First of all, we introduce a change of variables from McGehee [Mc]. This change avoids the singularity q=0 in (1). Later, we shall consider another change introduced by Lacomba and Simó [LS] in order to study the singularity $\|q\| = +\infty$.

(I.2) Collision manifold

McGehee's coordinates (r, θ, v, u) are defined by (for more details see Devaney ([D5]):

$$\begin{aligned}
r &= (q^t M q)^{\frac{1}{2}} \\
\theta &= \arctan(q_2/q_1)
\end{aligned} \tag{2}$$

Received by the editor February 29, 1984 and, in revised form, May 18, 1984.

1

$$u = r^{3/2} \dot{\theta}$$
$$v = r^{1/2} \dot{r} \qquad\qquad (2)$$
$$dt/d\tau = r^{3/2}$$

Then the Kepler problem is defined by,

$$r' = r\, v$$
$$v' = \tfrac{1}{2}v^2 + u^2 - 1 \qquad\qquad (3)$$
$$\theta' = u$$
$$u' = -\tfrac{1}{2}vu$$

where the prime indicates differentiation with respect to τ.

The energy relation and the angular momentum are,

$$\tfrac{1}{2}(u^2 + v^2) - 1 = rh, \quad \text{and} \qquad\qquad (4)$$
$$r^{\frac{1}{2}}\,\theta' = c, \qquad\qquad (5)$$

respectively. Note that the change (2) is not canonic, so system (3) is not Hamiltonian.

Now, the vector field (3) is analytic on the invariant boundary $r=0$, denoted by Λ and called <u>collision</u> <u>manifold</u>. The energy relation (4) shows that Λ is a two dimensional torus given by,

$$\Lambda = \{(r,v,\theta,u): r=0,\ \tfrac{1}{2}(u^2+v^2)=1,\ \theta\in S^1\}.$$

On Λ system (3) becomes,

$$v' = \tfrac{1}{2}u^2$$
$$\theta' = u \qquad\qquad (6)$$
$$u' = -\tfrac{1}{2}vu$$

and it has two circles $S^{\pm}$ of equilibrium points defined by $v = \pm\sqrt{2}$, $u=0$, $\theta\in S^1$. Of course, these points are also equilibrium points of (3).

Solutions on Λ move from the lower circle S^- to the upper one, S^+. Each rest point $(v = -\sqrt{2},\ \theta=\theta_o, 0)$ has associated a unique unstable manifold which is the stable manifold of the rest point $(v=\sqrt{2},\ \theta=\theta_o, 0)$.

If we consider the flow of (3), we must add the coordinate r. Then the flow on and near the collision manifold is given in Figure 1 (see Theorem 3.2 and Proposition 3.3 of [D2]).

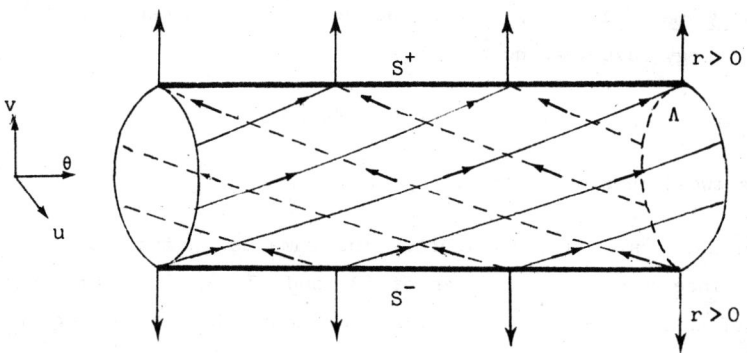

Figure 1. The flow on and near the collision manifold for the Kepler
         Problem.

Since Λ does not depend on the energy h, Λ lies on the boundary of
each energy level H=h.

(I.3) Infinity manifolds

Now, we shall study the singularity r=∞. From (4) we have rh+1 $\geq$ 0.
If h < 0, then the motion is bounded by the circle of zero velocity r=-1/h.
So, r can only reach the infinity when h $\geq$ 0.

In order to study the infinity it is necessary to treat the energy
levels h=0 and h > 0 separately.

First of all, we consider the case h=0. We take $\rho=r^{-1}$; then system
(3) becomes,

$$\rho' = -\rho v$$
$$v' = \tfrac{1}{2}u^2$$
$$\theta' = u \qquad\qquad\qquad (7)$$
$$u' = -\tfrac{1}{2}vu$$

The new energy relation and angular momentum are given by,

$$\tfrac{1}{2}(u^2+v^2) - 1 = 0, \qquad \text{and} \qquad\qquad (8)$$
$$\rho^{-\frac{1}{2}}\,\theta' = c \qquad\qquad\qquad (9)$$

From (7) $\rho=0$ is an invariant manifold under the flow. We shall call it _infinity_ _manifold_, N_o, and it appears as a boundary manifold glued to the zero energy level. We note that,

$$N_o = \{(\rho,v,\theta,u): \rho=0, \tfrac{1}{2}(u^2+v^2)=1, \theta \in s^1\}$$

is also a two-dimensional torus.

On N_o, system (7) is exactly the same as on the collision manifold. However, since the first equation of (3) and (7) are the same, with the exception of a sign, we get the flow $\rho(\tau)$ near N_o reversing the sense of the flow $r(\tau)$ near Λ, see Figure 1.

Now, we consider the case $h > 0$. We make the change, $\rho=r^{-1}$, $W=\rho^{\frac{1}{2}}v$, $U = \rho^{\frac{1}{2}}u$ and $d\tau/ds = \rho^{\frac{1}{2}}$. From (3), (4) and (5) it follows that,

$$\begin{aligned}
\rho' &= -\rho W \\
W' &= U^2-\rho \\
\theta' &= U \\
U' &= -WU
\end{aligned} \qquad (10)$$

where, now, the prime indicates differentiation with respect to s.

The energy relation and angular momentum go over to,

$$\tfrac{1}{2}(U^2+W^2)-\rho = h \qquad (11)$$

and

$$\rho^{-1}\theta' = c \qquad (12)$$

From (10) we have that $\rho=0$ is an invariant manifold under the flow, N_h, called the _infinity_ _manifold_, glued to the energy level H=h. That is,

$$N_h = \{(\rho,W,\theta,U): \rho=0, \tfrac{1}{2}(U^2+W^2) = h, \theta \in s^1\}.$$

By using the change $(W,\theta,U) = \frac{1}{2}(\overline{W},\overline{\theta},\overline{U})$ it is immediate that the expresions of the equations are the same than in the case N_o (of course, they are not equivalent because they are defined on different spaces).

(I.4) Summary on the singularities

We have made a "blow up" of the singularities and replaced them with invariant boundary manifolds. That is, system (1) has been extended analytically over the collision manifold Λ and over the infinity manifold N_h when $h \geq 0$.

Since we know the flow on Λ and N_h, we can understand the behaviour of the flow near them. It is easy to see that the flow is normally hyperbolic at Λ and N_h. Then, from [HPS] (see [D2]) we can obtain Table 1.

Invariant manifold V	Local unstable manifold at V	Local stable manifold at V
Λ	$S^1 \times \mathbb{R}$	$S^1 \times \mathbb{R}$
N_o	$\mathbb{R}^3$	$\mathbb{R}^3$
N_h	$\mathbb{R}^3$	$\mathbb{R}^3$

Table 1.

We denote by V^u (resp. V^s) the <u>unstable</u> (resp. <u>stable</u>) <u>manifold at</u> <u>V</u>. Note that Λ^u (resp. Λ^s) is formed by the ejection (resp. collision) orbits and N_h^u (resp. N_h^s) by the capture (resp. escape) orbits. These orbits are the unique ones such that their limit coordinates are the coordinates of the equilibrium points.

For an arbitrary system, the unstable and stable invariant manifolds can only be described locally. However, since system (1) is integrable we can compute these invariant manifolds globally (see (I.6)).

(I.5) Heteroclinic orbits

A solution $(r(\tau), v(\tau), \theta(\tau), u(\tau))$ of (3) such that $\theta(\tau)$ is constant is called a <u>homothetic orbit</u>. From (5), such orbits have angular momentum c=0, and from (3) and (4) they satisfy,

$$r' = rv,$$
$$v' = rh,$$
$$\theta = \text{constant},$$
$$u = 0, \quad \text{and}$$
$$\tfrac{1}{2}v^2 = rh + 1$$

then the phase portrait in the plane (r,v) is given in Figure 2 (see Figure 2 of [D5]). Its projection on the configuration space (r,θ) is shown in Figure 3.

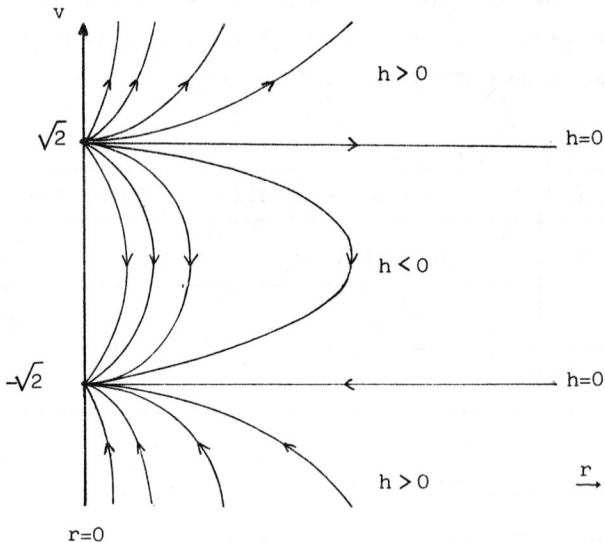

Figure 2. Homothetic orbits on the (r,v) plane.

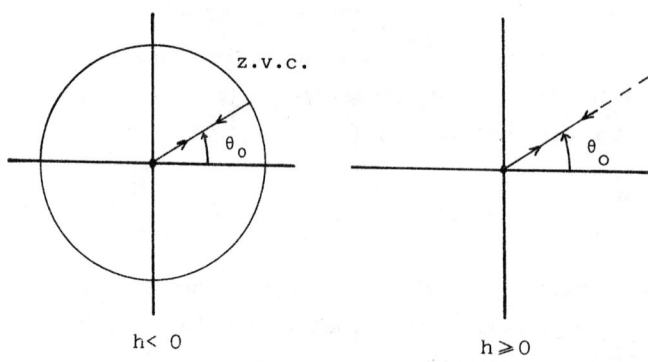

Figure 3. Homothetic orbits on the position plane (r,θ).
Here z.v.c. means the zero velocity curve.

Since the points $(r=0,\ v=\pm\sqrt{2},\ \theta=\text{constant},\ u=0)$ belong to Λ, then from Figure 2 we have that all homothetic orbits are collision or ejection orbits. Conversely, from (5) it follows that each collision or ejection orbit is a homothetic orbit. Then the invariant manifolds at Λ are formed by the homothetic orbits.

So, from Figure 2 and Table 1 it follows that

$$\Lambda_h^u = \Lambda_h^s \quad \text{if} \quad h < 0 \qquad \text{and}$$

$$\Lambda_h^u \cap \Lambda_h^s = \emptyset, \quad \Lambda_h^u \subsetneqq N_h^s \quad \text{and} \quad \Lambda_h^s \subsetneqq N_h^u \quad \text{if} \quad h \geq 0.$$

Hence, all the homothetic orbits are heteroclinic orbits. When $h < 0$, $h = 0$, $h > 0$ these orbits are called <u>elliptic</u>, <u>parabolic</u>, <u>hyperbolic</u> <u>collision</u> or <u>ejection</u> <u>orbits</u>, respectively.

(I.6) Global flow

Recall that the Hamiltonian system (1) is integrable, because the angular momentum C is another integral in involution with H. We consider the invariant sets,

$$I_c = \{(q,p): \ C(p,q) = c\},$$
$$I_h = \{(q,p): \ H(p,q) = h\},$$
$$I_{ch} = I_c \cap I_h.$$

If (c,h) is a regular value of the function (C,H), then I_{ch} is an invariant two-dimensional manifold. In this case, by using Liouville-Arnold's theorem(or merely, by elementary considerations) we have that I_{ch} is diffeomorphic to $S^1 \times S^1$, $S^1 \times \mathbb{R}$ or $\mathbb{R} \times \mathbb{R}$ (see Theorem 5.2.21 of [AM]). Furthermore, the flow on I_{ch} is a linear flow (see Theorem 5.2.23 of [AM]).

System (1) is a Hamiltonian system with symmetry. That is, the Lie Group S^1 acts diagonally as a transformation group on $(\mathbb{R}^2 - \{(0,0)\}) \times \mathbb{R}^2$, as a group of isometries with respect to the kinetic energy and leaves the potential energy V invariant (for more details see [S1,2]).

From Corollary 4.5 of [S2] it follows that S^1 also acts on I_{ch}. So, if (c,h) is a regular value then I_{ch} is diffeomorphic either to $S^1 \times S^1$ or $S^1 \times \mathbb{R}$. Of course, if I_{ch} is a compact two manifold then $I_{ch} \approx S^1 \times S^1$. In this case, by Theorem 5.2.23 of [AM], in action-angle variables the differential equations are linear on I_{ch}, and the orbits on I_{ch} are periodic or dense depending on whether the frequencies are rationally dependent or independent. However in this problem, if $I_{ch} = S^1 \times S^1$ all the orbits are periodic.

Now, we shall describe how the solutions foliate the sets I_{ch} and how these sets foliate the phase space of the Kepler problem.

From (4) and (5) we have that every solution of the Kepler problem satisfies,

$$(r')^2 = r(2hr^2 + 2r - c^2) \tag{13}$$

Figure 4 shows us the above curves on the plane (r,r'). Since these curves are symmetric with respect to the r-axis and $r > 0$, we have drawn them only on the first quadrant.

We obtain two different qualitative pictures for the curves (13) depending on whether $h < 0$ or $h \geqslant 0$. Furthermore, if $h < 0$, then the unique possible values of c belong to $[-(-2h)^{-1/2}, (-2h)^{-1/2}] = J_h$. Otherwise, all the values of c are possible.

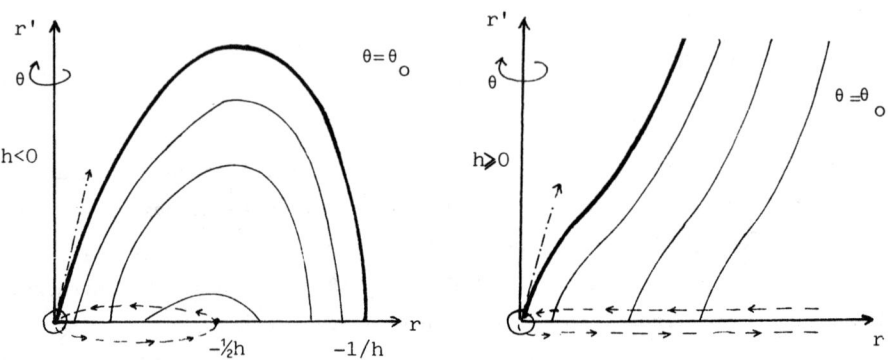

Figure 4. The curves $(r')^2 = r(2hr^2+2r-c^2)$. Here we denote by

> —————— the curve associated to $c=0$
>
> - - - - - → the intersections of I_{ch} with $(r,r',\theta=\theta_o)$ when c is moving in $[-(-\frac{1}{2}h)^{\frac{1}{2}}, (-1/2h)^{\frac{1}{2}}]$ if $h<0$ or in $\mathbb{R}$ if $h \geq 0$.
>
> - · - · - → the tangent vector to the curve $c=0$ at the point $r=0$.   It has slope equal to $\sqrt{2}$.
>
> $\odot$ the collision torus (see Figure 1).

Let h be a fixed value of energy. Then, we have,

$$I_h = \bigcup_{c \in J_h} I_{ch} \qquad \text{if} \quad h < 0$$

$$I_h = \bigcup_{c \in \mathbb{R}} I_{ch} \qquad \text{if} \quad h \geq 0$$

In order to describe I_h it is sufficient to take into account the three coordinates (r,v,θ). Since $r'=rv$ and θ do not appear in (13), from Figure 4 we obtain Table 2.

energy	angular momentum	topology of I_{ch}		I_{ch} is formed by
h < 0	$c = -(-2h)^{-\frac{1}{2}}$	S^1	(1)	one circular retrograde orbit
	$c \in (-(-2h)^{-\frac{1}{2}}, 0)$	$S^1 \times S^1$	(2)	a set of S^1 elliptic retrograde orbits
	$c = 0$	$S^1 \times \mathbb{R}$	(3)	a set of S^1 elliptic ejection-collision orbits $\equiv \Lambda_h^u = \Lambda_h^s$
	$c \in (C, (-2h)^{-\frac{1}{2}})$	$S^1 \times S^1$	(4)	a set of S^1 elliptic direct orbits
	$c = (-2h)^{-\frac{1}{2}}$	S^1	(5)	one circular direct orbit.
h = 0	$c \in (-\infty, 0)$	$S^1 \times \mathbb{R}$	(1)	a set of S^1 parabolic capture-escape retrograde orbits
	$c = 0$	2 copies of $S^1 \times \mathbb{R}$	(2)	$\Big\{$ a set of S^1 parabolic ejection-escape orbits $= \Lambda_o^u$ / a set of S^1 parabolic capture-collision orbits $= \Lambda_o^s$
	$c \in (C, \infty)$	$S^1 \times \mathbb{R}$	(3)	a set of S^1 parabolic capture-escape direct orbits
h > 0	$c \in (-\infty, 0)$	$S^1 \times \mathbb{R}$	(1)	a set of S^1 hyperbolic capture-escape retrogade orbits
	$c = 0$	2 copies of $S^1 \times \mathbb{R}$	(2)	$\Big\{$ a set of S^1 hyperbolic ejection-escape orbits $= \Lambda_h^u$ / a set of S^1 hyperbolic capture-collision orbits $= \Lambda_h^s$
	$c \in (C, \infty)$	$S^1 \times \mathbb{R}$	(3)	a set of S^1 hyperbolic capture-escape direct orbits

Table 2. The numbers correspond to the ones in Figure 6b and Figure 7b.

A solution contained in I_{ch} when $c \neq 0$ is called elliptic, parabolic or hyperbolic because its projection on the configuration plane is an ellipse, parabola or hyperbola, respectively. If $c=0$, the projection orbit is contained in a straight line, see Figure 3. A solution is called retrogade (resp. direct) if its projection on the configuration plane turns clockwise (resp. counterclockwise) around the origin.

Since S^1 acts on I_{ch}, we have that S^1 appears in every I_{ch} of Table 2. We note that $(c=\pm(-2h)^{-1/2},h)$ are the unique critical values of (C,H). It is well known that all the solutions when $h < 0$ and $c \neq 0$ are ellipses, thus every torus $S^1 \times S^1$ is formed by periodic orbits.

In Figures 5 and 6 it is represented the invariant set I_h with $h<0$.

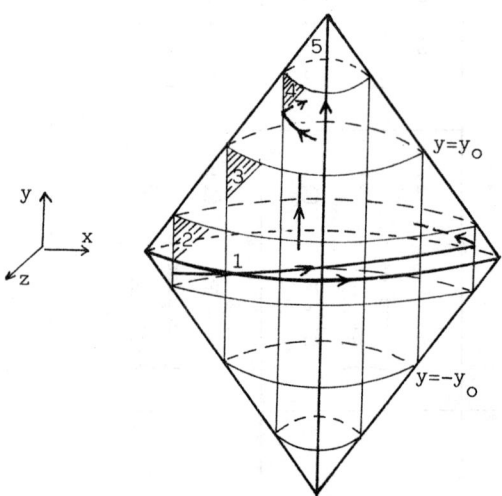

Figure 5. In this picture we identify the point (x,y,z) with $(x,-y,z)$ for all $y \neq y_0$ if it is in the boundary. Then we obtain the invariant set $I_h = \bigcup_c I_{ch}$, when $h<0$. The numbers $1,2,3,4,5$ correspond to the numbers of Table 2.

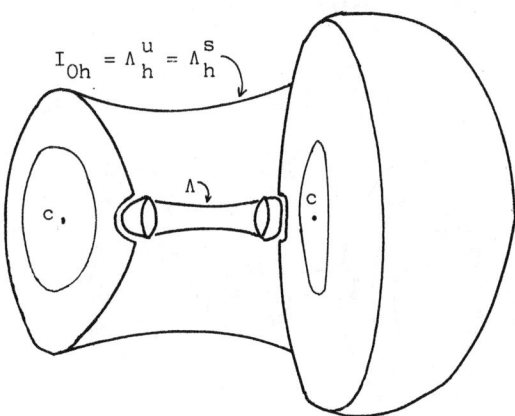

Figure 6.a. The invariant set $\bigcup_{c \geqslant 0} I_{ch}$ or $\bigcup_{c \leqslant 0} I_{ch}$ for $h<0$. It is obtained by rotating Figure 4 around the r'-axis and by gluing the collision manifold. Actually, we have only drawn either the retrograde orbits or the direct ones. Here c denotes the circular orbit.

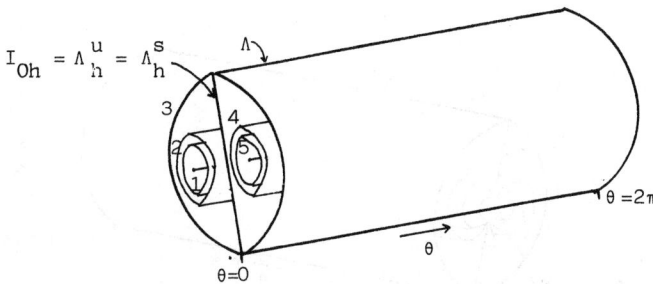

Figure 6.b. The set $I_h = \bigcup_c I_{ch}$, $h<0$ obtained by gluing the collision manifold Λ to Figure 5.

From Figure 5, if we do not regularize collisions then the topology of I_h is $S^3 \setminus S^1$. If we add the collision manifold, which corresponds to glue a torus with one circle identified to the circle $y=y_o$, and another one to the circle $y=-y_o$, then the topology of I_h becomes a closed three-dimensional solid torus, see Figures 6.

Note that the topology of I_h depends on the choosen coordinates. It is well known that using other coordinates $I_h \approx P^3$.

Now Figures 7 show the invariant set I_h with $h \geqslant 0$.

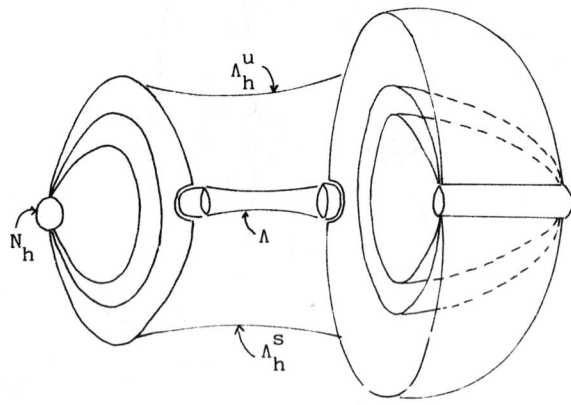

<u>Figure 7.a</u>. The invariant set $\underset{c \geqslant o}{U} I_{ch}$ or $\underset{c \leqslant o}{U} I_{ch}$ for $h \geqslant 0$. It is obtained
by rotating Figure 4 around the r'-axis and gluing the collision
and the infinity manifolds. In fact, we have only drawn either
the retrograde orbits or the direct ones.

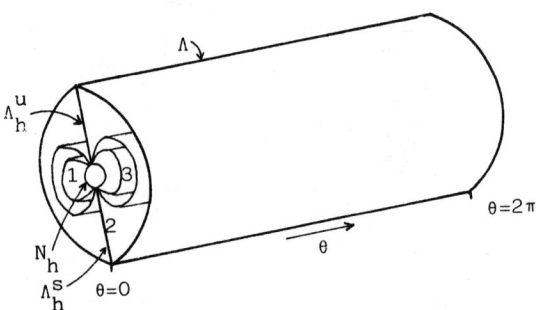

<u>Figure 7.b</u>. The invariant set $I_h = \underset{c}{U} I_{ch}$ for $h \geqslant 0$. The numbers 1,2,3 corres-
pont to the numbers of Table 2. Here, we have glued to I_h the colli_
sion manifold Λ and the infinity manifold N_h.

If $c \neq 0$ it follows that each elliptic (resp. parabolic or hyperbolic)
solution cuts r'=0 exactly in two (resp. one) points, see Figure 8.a (resp.
Figure 9.a). If c=0, then each elliptic (resp. parabolic or hyperbolic) solution cuts r'=0 exactly in one (resp. zero) point, see Figure 8.b (resp. Figure 9.b).

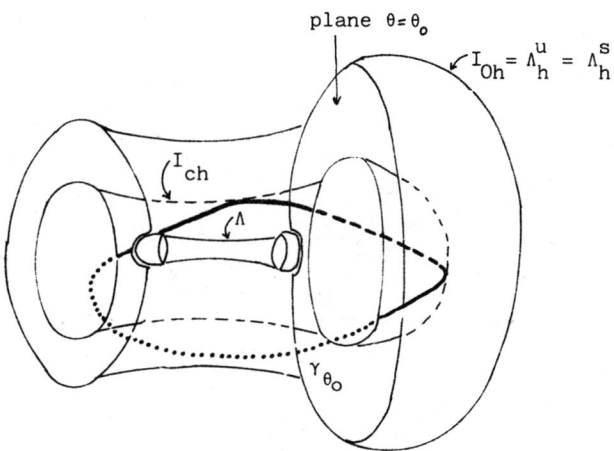

Figure 8.a. An elliptic orbit γ_{θ_0} on the two-dimensional torus I_{ch} for $h < 0$ and $c \neq 0$. This orbit has its apocenter in $\theta = \theta_0$ and its pericenter in $\theta = \theta_0 + \pi$.

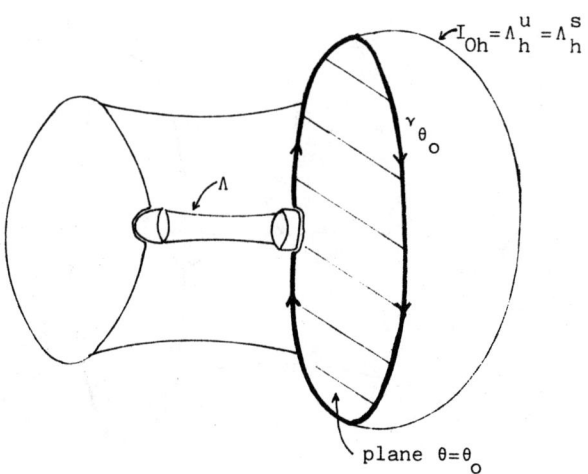

Figure 8.b. An elliptic ejection-collision orbit γ_{θ_0} (homothetic orbit) on I_{oh} for $h < 0$ and $c = 0$. Note that γ_{θ_0} is contained into the plane $\theta = \theta_0$.

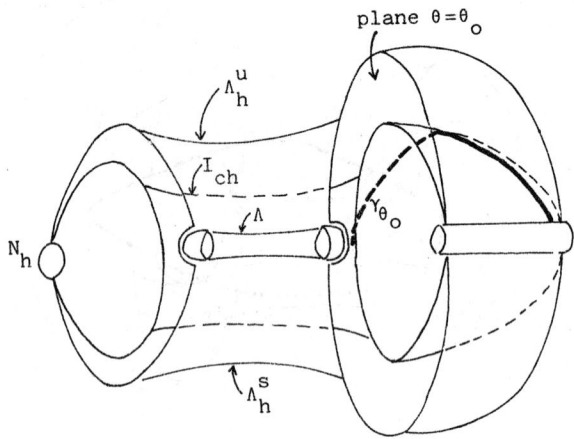

<u>Figure 9.a.</u> A parabolic (resp. hyperbolic) capture-escape orbit on I_{ch} for
h=0 (resp. h>0) and c≠0. This orbit has its pericenter in $\theta=\theta_o$ and
we have only drawn a half part of it.

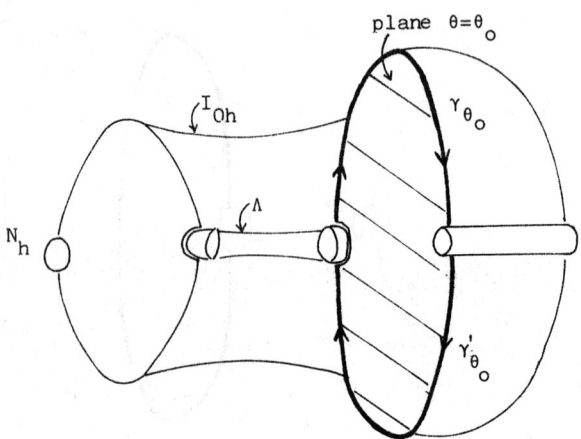

<u>Figure 9.b.</u> A parabolic (resp. hyperbolic) ejection-escape orbit γ_{θ_o} and a
parabolic (resp. hyperbolic) capture-collision orbit γ'_{θ_o} for
h=0 (resp. h>0) and c=0. Note that γ_{θ_o} and γ'_{θ_o} are contained into
the plane $\theta=\theta_o$.

(I.7) Poincaré map

In order to define a Poincaré map in the energy level H=h it is necessary to take $h < 0$.

Let $P = \{(\theta,u): v=0 \text{ and } H=h\} = \{(\theta,u): u \in [-\sqrt{2},\sqrt{2}], \theta \in S^1\}$.
We note that the two circular orbits are contained in P, see Figure 10. The retrogade (resp. direct) circular orbit is the circle $u=1$ (resp. $u=-1$).

The torus I_{ch} cuts P in the circles

$$u_1 = \frac{c}{|c|} [1-(1+2hc^2)^{\frac{1}{2}}]^{\frac{1}{2}}$$

and

$$u_2 = \frac{c}{|c|} [1+(1+2hc^2)^{\frac{1}{2}}]^{\frac{1}{2}},$$

and the boundary of P given by the circles $u=\pm\sqrt{2}$ belongs to the collision manifold Λ .   The circle $u=0$ is the zero velocity curve, which is only reached by the homothetic orbits (elliptic ejection-collision orbits).

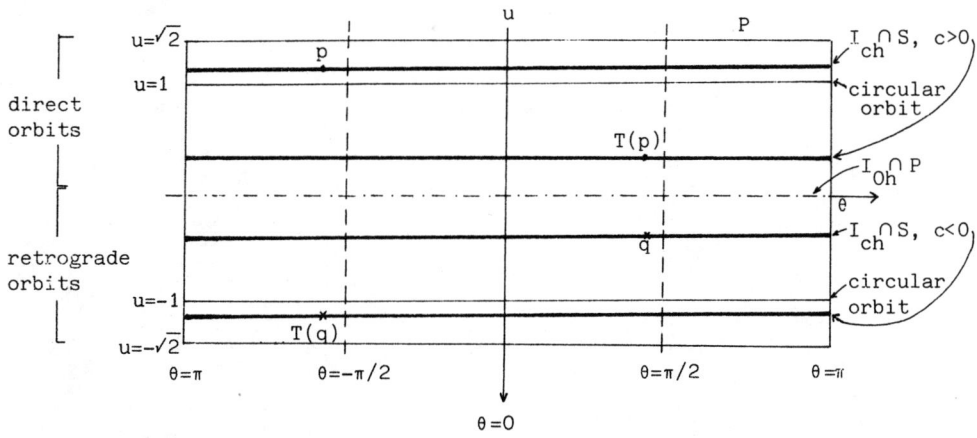

Figure 10. The Poincaré map T on the four rings of S.

We consider the following surface of section,

$$S = \{(\theta,u) \in P: u \neq \pm\sqrt{2}, u \neq \pm 1, u \neq 0\}.$$

For any point $(\theta,u) \in S$ we denote by (θ',u') the point at which the orbit that passes through $(r=(u^2/2 -1)/h, v=0, \theta, u)$ first meets S. The transformation

J. CASASAYAS & J. LLIBRE

which carries (θ, u) to (θ', u') defines a Poincaré map $T: S \longrightarrow S$ given by the equations,

$$(\theta', u') = \begin{cases} (\theta+\pi, \; [u^2+2(1+2hc^2)^{1/2}]^{1/2} \cdot u/|u|) & \text{if} \quad 0 < |u| < 1 \\[2ex] (\theta+\pi, \; [u^2-2(1+2hc^2)^{1/2}]^{1/2} \cdot u/|u|) & \text{if} \quad |u| > 1 \end{cases}$$

Note that T^2 = identity.

II. THE ANISOTROPIC KEPLER PROBLEM

(II.1) Formulation.

We consider the Hamiltonian,

$$H'(q,p) = (p^t M^{-1} p)/2 + V(q)$$

where H' is defined on $(\mathbb{R}^2 \setminus \{0\}) \times \mathbb{R}^2$, the mass matrix $\bar{M}^{-1} = \begin{pmatrix} \mu & 0 \\ 0 & 1 \end{pmatrix}$, and the potential energy $V(q) = -1/\|q\|$.

The goal of study the anisotropic Kepler problem is to describe the solutions of the Hamiltonian system associated to H',

$$\dot{q} = M^{-1} p$$
$$\dot{p} = -\text{grad } V(q)$$

They are a one parameter family of Hamiltonian systems depending analytically on the parameter $\mu \geq 1$. This system describes the Kepler problem when $\mu = 1$ (see Chapter I), and the case $\mu > 1$ corresponds to consider the q_2-axis as the "heavy axis".

From now on, we shall consider the Hamiltonian system,

$$\dot{q}_1 = p_1$$
$$\dot{q}_2 = p_2$$
$$\dot{p}_1 = -\mu q_1 (\mu q_1^2 + q_2^2)^{-3/2}$$
$$\dot{p}_2 = - q_2 (\mu q_1^2 + q_2^2)^{-3/2}$$

associated to the Hamiltonian,

$$H(q,p) = \|p\|^2/2 + V(q)$$

where $V(q) = -(\mu q_1^2 + q_2^2)^{-1/2}$ and H is defined on $(\mathbb{R}^2 \setminus \{0\}) \times \mathbb{R}^2$. We remark that the Hamiltonian systems associated to H and H' are equivalent and the angular momentum, $C(q,p) = q \wedge p$, is an integral if and only if $\mu = 1$.

The case $\mu \in [1, +\infty)$ that we shall study includes the case $\mu \in (0,1]$ by using the change: $\bar{q}_1 = q_2$, $\bar{q}_2 = q_1$, $\bar{p}_1 = \mu^{1/4} p_2$, $\bar{p}_2 = \mu^{1/4} p_1$ and $dt = \mu^{1/4} ds$.

17

If we have the energy level H=h≠0 then, the change of coordinates :
$\bar{q}_i = |h| q_i$, $\bar{p}_i = |h|^{-1/2} p_i$ for i=1,2 and, hence dt=$|h|^{-3/2}$ds, carries H=h to H=1 or
H=-1 according as h>0 or h<0 respectively. So, it is sufficient to study the
energy levels H=1, H=0 and H=-1.

(II.2) Symmetries.

We consider a 2n-dimensional manifold M together with a diffeomorphism R
of M satisfying,
(1) R^2 = identity   and
(2) dim (Fix(R))=n
then, R is called a reversing involution. A smooth vector field X on M is called
R-reversible if DR(X)= -X∘R; for more details on reversible systems see [D1] .

It is easy to verify that the anisotropic Kepler problem (1) is S_i-rever-
sible for i=0,1,2 where,

$$S_0(q_1, q_2, p_1, p_2) = (q_1, q_2, -p_1, -p_2),$$
$$S_1(q_1, q_2, p_1, p_2) = (q_1, -q_2, -p_1, p_2),$$
$$S_2(q_1, q_2, p_1, p_2) = (-q_1, q_2, p_1, -p_2).$$

This means that if $\gamma(t) = (q_1(t), q_2(t), p_1(t), p_2(t))$ is a solution of the
anisotropic Kepler problem such that $\gamma(0)$ belongs to Fix(S_0), Fix(S_1) or
Fix(S_2), then $(q_1(-t), q_2(-t), -p_1(-t), -p_2(-t))$, $(q_1(-t), -q_2(-t), -p_1(-t), p_2(-t))$,
or $(-q_1(-t), q_2(-t), p_1(-t), -p_2(-t))$ is, respectively, a solution.

The symmetry S_0 is the usual symmetry with respect to the zero velocity
curve, which is presented by all the Hamiltonian systems where the Hamiltonian
can be writen as kinetic energy, $(p^t M^{-1} p)/2$, plus potential energy, V(q).

A plane in the phase space is called invariant plane if and only if every
orbit which has a point in the plane is contained in it.

Let V_q be the gradient and V_{qq} be the Hessian of the potential V. Set
T= $-J V_{qq} J V_q$, where J =$\begin{pmatrix} 0 & 1 \\ -1 & 0 \end{pmatrix}$. By Lemma 2.1 of [CPR], the irreductible factors
of degree 1 of the equation< T,JV_q> =0 are the projections of the invariant pla-
nes on the configuration plane. Here, <,> denotes the Euclidean inner
product.

If $V(q) = -(\mu q_1^2 + q_2^2)^{-1/2}$ then we have, $<T,JV_q> = \mu(1-\mu)q_1 q_2 (\mu q_1^2 + q_2^2)^{-9/2}$. Therefore, the unique invariant planes of (1) are,

$$\pi_1 = \{(0,q_2,0,p_2): (q_2,p_2) \in (\mathbb{R}\setminus\{0\}) \times \mathbb{R}\},$$
$$\pi_2 = \{(q_1,0,p_1,0): (q_1,p_1) \in (\mathbb{R}\setminus\{0\}) \times \mathbb{R}\},$$

where for $i=1,2$, π_i is invariant under the symmetry S_j for $j=0,1,2$. In short we have proved the following proposition.

PROPOSITION 1. _The anisotropic Kepler problem has only two invariant planes,_ π_1 _and_ π_2 .

Of course, the flow restricted to an invariant plane is given by a Hamiltonian system of degree 1. Then the flow on π_1 is described in Figure 1. In a similar way we can obtain the flow on π_2.

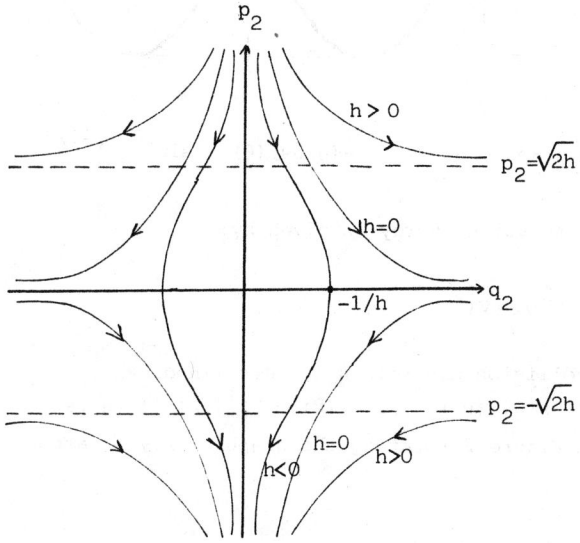

Figure 1. The flow on the plane π_1 given by the Hamiltonian $H = p_2^2/2 - 1/|q_2|$.

(II.3) The Collision Manifold.

From I.(2) the Hamiltonian system associated to H becomes:

$$r' = rv$$
$$v' = v^2/2 + u^2 + V(\theta)$$
$$\theta' = u$$
$$u' = -vu/2 - V'(\theta)$$

(2)

where the prime on the left part of these equations indicates differentiation with respect to τ, $V'(\theta)=dV(\theta)/d\theta$, and $V(\theta)= -(\mu\cos^2\theta + \sin^2\theta)^{-1/2}$. The graphic of $V(\theta)$ is given in Figure 2.

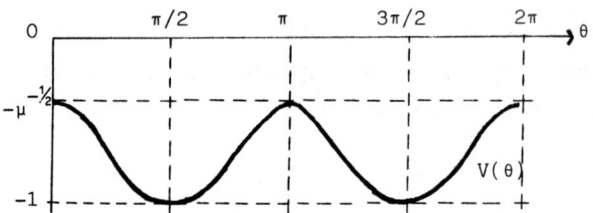

Figure 2. The graphic of $V(\theta)= -(\mu\cos^2(\theta) + \sin^2(\theta))^{-1/2}$.

Now, the relation energy is given by:

$$rh = (u^2+v^2)/2+V(\theta)$$

(3)

The collision manifold Λ is determined by
$\Lambda= \{ (r,v,\theta,u): r=0, (u^2+v^2)/2=-V(\theta), \theta \in S^1 \}$. It is a two dimensional torus obtained from Figure 2 rotating the graphic of $V(\theta)$ around the θ-axis, see Figure 3.

Since $r'=0$ on Λ, we have that Λ is invariant by the flow (2).

For $\mu =1$ system (2) has two circles of equilibrium points on Λ (see (I.2)). For $\mu \neq 1$ each of these circles breaks up into four equilibrium points. Devaney in [D2] proved the following proposition.

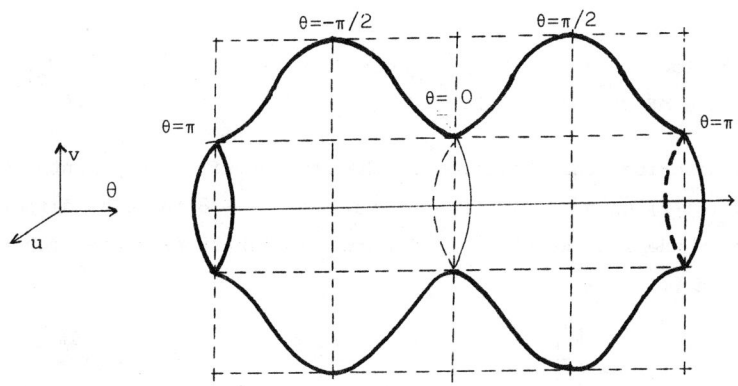

Figure 3. The collision manifold Λ.

PROPOSITION 2. (i) For every value $h \in R$ and $\mu \in (1, \infty)$, the equilibrium points of system (2) lie on Λ and they are given by:

$$(r=0, \quad v=\pm (-2V(\theta_o))^{1/2}, \quad \theta_o = -\pi/2, \; 0, \; \pi/2, \pi, \quad u=0) = p^{\pm}(\theta_o)$$

(ii) The eigenvalues and the dimensions of stable and unstable invariant manifolds associated to equilibrium points of system (2) are given in Table 1.

Equilibrium point	Characteristic exponents on Λ	off Λ	Dimensions of W^u, W^s on Λ	on $H=h$	Type on Λ
$p^+(0)$ $p^+(\pi)$	$A(-1 \pm (9-8\mu^{-1})^{1/2})$	$4A$	W^u 1 W^s 1	2 1	saddle
$p^+(\pi/2)$ $p^+(-\pi/2)$	$B(-1 \pm (9-8\mu)^{1/2})$	$4B$	W^u 0 W^s 2	1 2	sink
$p^-(0)$ $p^-(\pi)$	$A(1 \pm (9-8\mu^{-1})^{1/2})$	$-4A$	W^u 1 W^s 1	1 2	saddle
$p^-(\pi/2)$ $p^-(-\pi/2)$	$B(1 \pm (9-8\mu)^{1/2})$	$-4B$	W^u 2 W^s 0	2 1	source

Table 1. Here we use the notation $A=2^{-3/2}\mu^{-1/4}$ and $B=2^{-3/2}$.

COROLLARY 3. *If μ>9/8, then all the sinks and sources on Λ have characteristic exponents with the imaginary part different from zero. That is, they are spiral sinks and spiral sources.*

In what follows we denote by $p^{\pm}(\theta_o)=(r=0,\ v=\pm(-2V(\theta_o))^{1/2},\ \theta=\theta_{\hat{o}},\ u=0)$ the eight rest points of Proposition 2.

Figure 4 describes the evolution of the characteristic exponents (c.e.) for $p^{+}(\pi/2)$ and $p^{+}(-\pi/2)$ when μ goes from 1 to ∞. For μ> 9/8 the c.e. different from $2^{1/2}$ are on the line $Re=-2^{-3/2}$ and their imaginary part goes monotonously to infinity with μ.

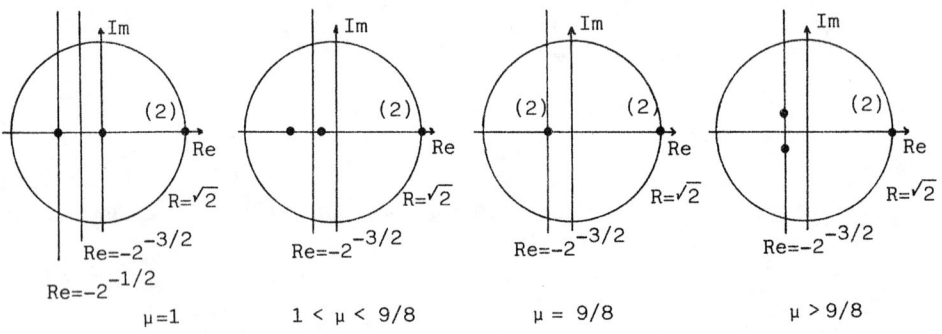

Figure 4. The evolution of characteristic exponents for $p^{+}(\pm\pi/2)$

Figure 5 describes the evolution of the c.e. for $p^{+}(0)$ and $p^{+}(\pi)$ when μ goes from 1 to ∞ . The behaviour of these c.e. is the following one: a) The c.e. equal to $-2^{-1/2}$ for μ=1 decreases when μ is increased reaching a minimum for $μ= 144/(53+\sqrt{217}\ )$ and after it increases monotonously to 0. The c.e. equal to 0 for μ=1 increases until $μ=144/(53-\sqrt{217}\ )$ and after it decreases monotonously to 0 when μ goes to infinity.

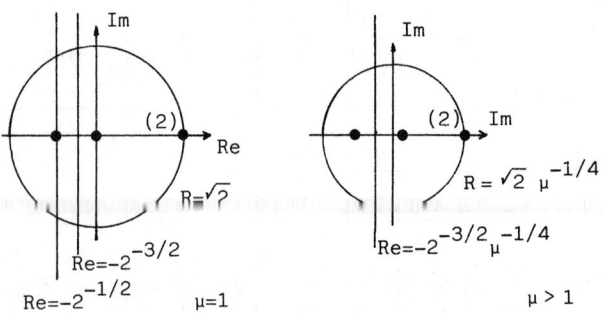

Figure 5. The evolution of characteristic exponents for $p^{+}(0)$ and $p^{+}(\pi)$.

REMARK 1. *We note that a solution can only reach (resp. leave)Λ through a stable (resp. unstable) invariant manifold of an equilibrium point on Λ.*

(II.4) The Infinity Manifold.

From (3) we have $rh-V(\theta)\geqslant 0$. If $h<0$, then the motion is bounded by the ellipse of zero velocity $r=V(\theta)/h$. So, r can only reach the infinity when $h\geqslant 0$. Again we shall study the cases $h=0$ and $h>0$ separately.

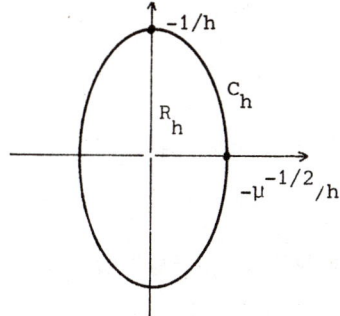

Figure 6. The zero velocity curve C_h and the region R_h for $h<0$.

First we consider the case $h=0$. If we introduce the change $\rho=r^{-1}$ then the equations (2) become:

$$\begin{aligned}
\rho' &= -\rho v \\
v' &= u^2/2 \\
\theta' &= u \\
u' &= -vu/2 - V'(\theta)
\end{aligned} \tag{4}$$

where we have used the energy relation $(u^2+v^2)/2 + V(\theta) = 0$. The manifold $\rho=0$ is invariant under the flow given by (4), and it is called the <u>infinity manifold</u> N_0. That is, $N_0 = \{(\rho,v,\theta,u): \rho=0, (u^2+v^2)/2=-V(\theta), \theta\in S^1\}$. So, N_0 is defined by the same equations as Λ (see (II.3)).

Now, we consider the case $h>0$. We make the change $\rho=r^{-1}$, $W=\rho^{1/2}v$, $U=\rho^{1/2}u$ and $d\tau/ds= \rho^{1/2}$. From (2) we obtain:

$$\begin{aligned}
\rho' &= -\rho W \\
W' &= U^2 + \rho V(\theta) \\
\theta' &= U \\
U' &= -WU - \rho V'(\theta)
\end{aligned} \tag{5}$$

where the prime on the left side indicates differentiation with respect s and $V'(\theta) = dV(\theta)/d\theta$.

From (3) the energy relation becomes:

$$h = (U^2 + W^2)/2 - \rho V(\theta) \qquad (6)$$

Again, $\rho = 0$ is an invariant manifold under the flow given by (5), denoted by N_h, and called the <u>infinity manifold</u> at <u>the energy level</u> h. So, $N_h = \{(\rho, W, \theta, U): \rho = 0, h = (U^2 + W^2)/2, \theta \in S^1\}$. Hence, N_h is equivalent to the collision manifold for the Kepler problem, see (I.2).

In short we have the following proposition.

PROPOSITION 4. *(i) If $\mu > 1$ and $h=0$, then the equilibrium points for system (4) are given by:*

$$P_-^+(\theta_o) = (\rho=0,\; v=\pm(-2V(\theta_o))^{1/2},\; \theta_o = -\pi/2, 0, \pi/2, \pi,\; u=0)$$

(ii) If $\mu > 1$ and $h > 0$, then the equilibrium points for system (5) are given by:

$$P_-^+(\theta) = (\rho=0,\; W=\pm(2h)^{1/2},\; \theta \in S^1,\; U=0)$$

(iii) The eigenvalues of stable and unstable invariant manifolds associated to the equilibrium points of (i) and (ii) are given in Table 2.

Equilibrium point	Characteristic exponents on N_o	off N_o	Dimensions of W^u, W^s on N_o	on $H=h$
$P^+(0)$ $P^+(\pi)$	$A(-1\pm(9-8\mu^{-1})^{1/2})$	$-4A$	W^u 1	1
			W^s 1	2
$P^+(\pi/2)$ $P^+(-\pi/2)$	$B(-1\pm(9-8\mu)^{1/2})$	$-4B$	W^u 0	0
			W^s 2	3
$P^-(0)$ $P^-(\pi)$	$A(1\pm(9-8\mu^{-1})^{1/2})$	$4A$	W^u 1	2
			W^s 1	1
$P^-(\pi/2)$ $P^-(-\pi/2)$	$B(1\pm(9-8\mu)^{1/2})$	$4B$	W^u 2	3
			W^s 0	0

Case h=0. Here we use $A = 2^{-3/2}\mu^{-1/4}$ and $B = 2^{-3/2}$.

Equilibrium point	Characteristic exponents		Dimensions of W^u, W^s		
	on N_h	off N_h		on N_h	on $H=h$
$P^+(\theta)$	$0, -(2h)^{1/2}$	$-(2h)^{1/2}$	W^u	0	0
			W^s	1	2
$P^-(\theta)$	$0, (2h)^{1/2}$	$(2h)^{1/2}$	W^u	1	2
			W^s	0	0

Case $h>0$.

Table 2.

We note that Table 2 for $h=0$ follows from the Proposition.

(II.5) Invariant manifolds I_h.

We fix a negative value of the energy h. The zero velocity curves C_h, $\bar{C}_h$; the Hill's regions R_h, $\bar{R}_h$ and the invariant manifolds I_h, $\bar{I}_h$ are given by:

$$C_h = \{q: h-V(q)=0\},$$
$$\bar{C}_h = \{(r,\theta): rh - V(\theta)=0\},$$
$$R_h = \{q: h-V(q)\geqslant 0\},$$
$$\bar{R}_h = \{(r,\theta): rh - V(\theta)\geqslant 0\},$$
$$I_h = \{(q,p): h-V(q)\geqslant 0 \text{ and } \|p\|^2 = 2(h-V(q))\},$$
$$\bar{I}_h = \{(r,v,\theta,u): rh-V(\theta)\geqslant 0 \text{ and } u^2+v^2 = 2(rh-V(\theta))\}.$$

Note that $C_h=\bar{C}_h$, $R_h \subsetneq \bar{R}_h$ and $I_h \subsetneq \bar{I}_h$ since $\bar{R}_h$ and $\bar{I}_h$ take into account the collision manifold.

When $h\geqslant 0$ we denote by $\bar{I}_h$ (resp. $\bar{R}_h$) the manifold I_h (resp. R_h) together with the collision and infinity manifold.

It is clear that $R_h = \mathbb{R}^2 \setminus \{(0,0)\}$ and $\bar{R}_h = [0,\infty]\times S^1$ if $h\geqslant 0$, and R_h is as in Figure 6 when $h<0$. That is, R_h (resp. $\bar{R}_h$) is topologically a punctured open disk (resp. closed annulus) when $h\geqslant 0$ and a punctured closed disk (resp. closed annulus) when $h<0$.

_LEMMA 5. (i) (see Proposition 1.1 and 2.1 of [D2]). If h<0 then I_h is diffeo-
morphic to an open solid torus and $\bar{I}_h$ is diffeomorphic to a solid torus with
boundary. The added boundary is the collision manifold Λ._

_(ii) If h⩾0, then I_h is diffeomorphic to ($\mathbb{R}^2 \setminus \{(0,0)\}) \times S^1$ (i.e, an open toroi-
dal annulus), $\bar{I}_h$ is diffeomorphic to a closed toroidal annulus and the inner
(resp. outer) boundary of this manifold is the collision manifold Λ(resp. the
infinity manifold N_h)._

(II.6) Heteroclinic orbits.

We recall that a solution $(r(\tau), v(\tau), \theta(\tau), u(\tau))$ of (2) is homothetic
when $\theta(\tau)$ is constant.

For the Kepler problem ($\mu = 1$) we know (see Figures I.8b and I.9b) that the
invariant manifolds (cylinders) I_{0h} are formed by homothetic orbits. These orbit
are ejection-collision or ejection-escape and capture-collision according as
h<0 or h⩾0.

In (II.4) we have seen that the two circles of equilibrium points for $\mu = 1$
break into eight equilibrium points when $\mu > 1$. This is due to the fact that the
critical points of potential energy $V(\theta)$ are all the values of $\theta \in S^1$ when $\mu = 1$
and only the values $\theta_o = -\pi/2, 0, \pi/2, \pi$ when $\mu > 1$. For the same reason we shall
see that each cylinder of homothetic orbits for $\mu = 1$ breaks into four homothetic
orbits at $\theta_o = -\pi/2, 0, \pi/2, \pi$.

From (2) and (3) the homothetic orbit at $\theta = \theta_0$ satisfies:

$$
\begin{aligned}
r' &= rv \\
v' &= rh \\
V'(\theta_o) &= 0 \\
u &= 0 \quad \text{and} \\
v^2/2 &= rh - V(\theta)
\end{aligned}
\tag{7}
$$

Then $\theta_0 = -\pi/2, 0, \pi/2, \pi$. We denote by $\gamma_h(\theta_o)$ the homothetic orbit at $\theta = \theta_o$ in
the energy level H=h. The phase portrait in the plane $(r, v, \theta = \theta_0, u = 0)$ is
given as in Figure I.2. This phase portrait is equivalent to the corresponding
one of Figure 1.

Since the points $(r=0; v=\overset{+}{-}(-2V(\theta_o))^{1/2}; \theta_o=-\pi/2,0,\pi/2,\pi; u=0)$ are the equilibrium points belonging to Λ, $(\rho=0, v=\overset{+}{-}(-2V(\theta_o))^{1/2}; \theta_o=-\pi/2,0,\pi/2,\pi; u=0)$ the equilibrium points of N_o and $(\rho=0, V=\overset{+}{-}(2h)^{1/2}; \theta_o=-\pi/2,0,\pi/2,\pi; U=0)$ the equilibrium points of N_h with $h>0$, we have that all the homothetic orbits are ejection-collision if $h<0$ and ejection-escape or capture-collision if $h>0$.

While for $\mu=1$ there is an equivalence between homothetic orbits and collision or ejection orbits, Tables 1 and 2 prove that this is not true for $\mu>1$.

In Figure 7 are represented the four (respectively eight) homothetic orbits on the phase space $\bar{I}_h$ for $h<0$ (resp. $h>0$) obtained as in (I.5).

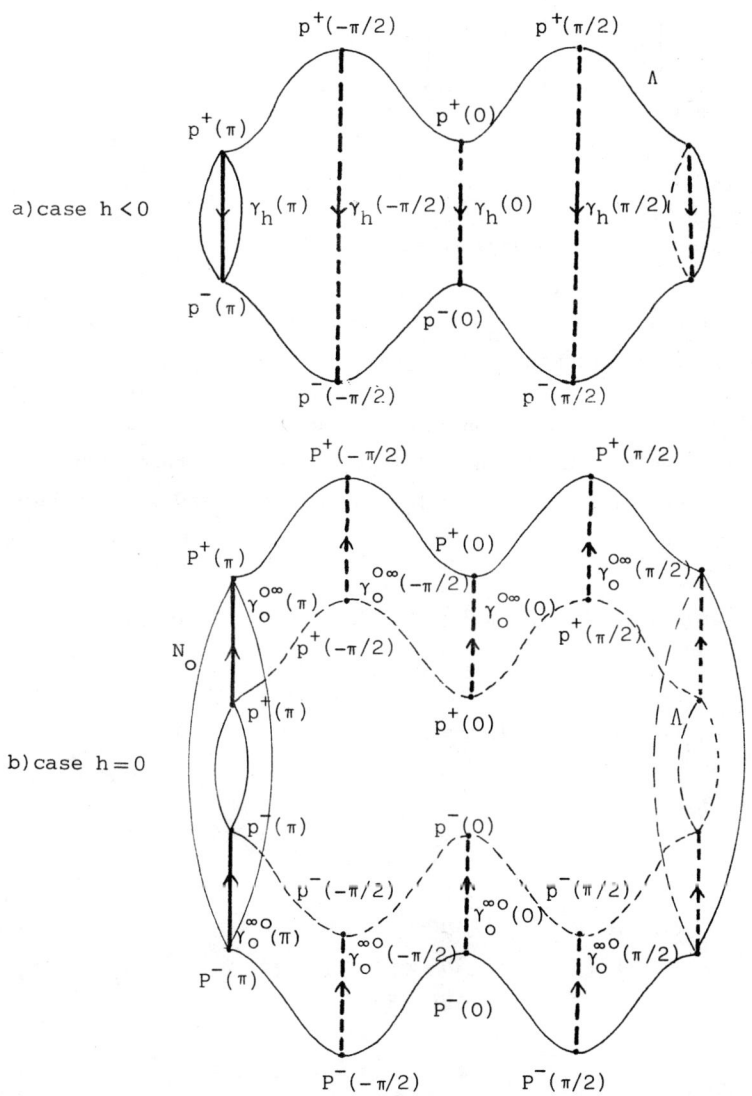

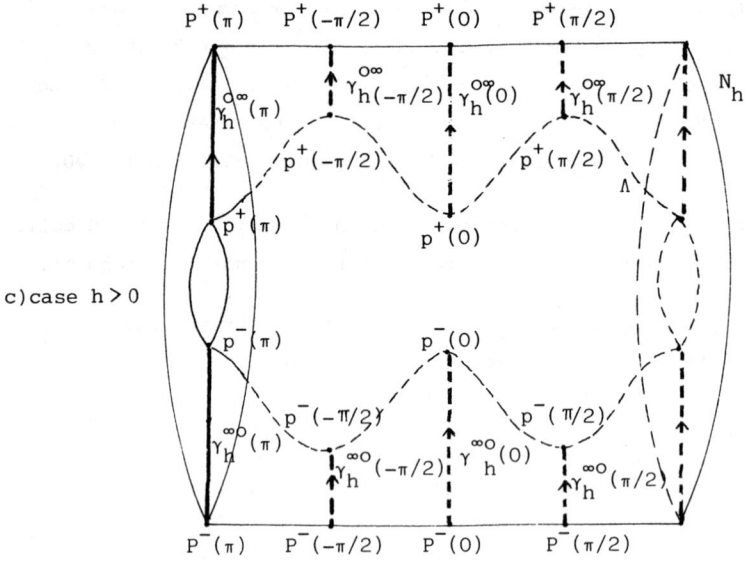

Figure 7. Homothetic orbits on the energy level $\bar{I}_h$.

Let p be an equilibrium point in $\bar{I}_h$. We denote by $W^u(p)$ (resp. $W^s(p)$) the unstable (resp. stable) invariant manifold of p restricted to I_h.

By Propositions 2 and 4 the collision (resp. ejection) orbits form the stable (resp. unstable) manifold of Λ, and the escape (resp. capture) orbits form the stable (resp. unstable) manifold of N_h. From Table 1 and 2 the following theorem holds.

THEOREM 6. (i) (see Devaney [D2]). If $\mu>1$ and $h<0$, then we have,

$$\gamma_h(\theta_o) = W^u(p^+(\theta_o)) = W^s(p^-(\theta_o)) \text{ when } \theta_o = -\pi/2, \pi/2$$
$$\gamma_h(\theta_o) \subsetneq W^u(p^+(\theta_o)) \cap W^s(p^-(\theta_o)) \text{ when } \theta_o = 0, \pi$$

(ii) If $\mu>1$ and $h\geq0$, then we have,

$$\gamma_h^{o\infty}(\theta_o) = W^u(p^+(\theta_o)) \subsetneq W^s(P^+(\theta_o)) \text{ for } \theta_o = -\pi/2, \pi/2$$
$$\gamma_h^{\infty o}(\theta_o) = W^s(p^-(\theta_o)) \subsetneq W^u(P^-(\theta_o)) \text{ for } \theta_o = -\pi/2, \pi/2$$
$$\gamma_h^{o\infty}(\theta_o) \subsetneq W^u(p^+(\theta_o)) \cap W^s(P^+(\theta_o)) \text{ for } \theta_o = 0, \pi$$
$$\gamma_h^{\infty o}(\theta_o) \subsetneq W^s(p^-(\theta_o)) \cap W^u(P^-(\theta_o)) \text{ for } \theta_o = 0, \pi$$

where $\gamma^{0\infty}$ (resp. $\gamma^{\infty 0}$) means a homothetic orbit of ejection-escape (resp. capture-collision).

Two submanifolds M_1 and M_2 of a manifold M <u>meet</u> <u>transversally</u> at a point $x \in M_1 \cap M_2$ if $T_x(M_1) + T_x(M_2) = T_x(M)$, where $T_x N$ is the tangent space to N at the point x. We say that M_1 meets M_2 transversally at $M_3 \subset M_1 \cap M_2$ if for all $x \in M_3$, M_1 and M_2 meet transversally at x.

From Theorem 6 and Table 1 it follows that a necessary condition in order that $\gamma_h(\theta_0)$ is a transversal homothetic orbit, is that $\theta_0 = 0, \pi$. In fact, Devaney in [D4] proved that this condition is sufficient. Now, we give a different proof using ideas of [CLL] and [LLS] .

<u>THEOREM 7</u>. *If $\theta_0 = 0, \pi$ then $W^u(p^+(\theta_0))$ meets transversally $W^s(p^-(\theta_0))$ along the homothetic orbit $\gamma_h(\theta_0)$ on $\bar{I}_h$ with $h < 0$.*

<u>Proof</u>. First of all, we shall use variational equations in order to study the tangent space to $W^u(p^+(\theta_0))$ along $\gamma_h(\theta_0)$. The symmetry S_0 provides us with the corresponding properties for the tangent space to $W^s(p^-(\theta_0))$.

From (2) we get the variational matrix along $\gamma_h(\theta_0)$:

$$
E = \begin{pmatrix}
v(\tau) & r(\tau) & 0 & 0 \\
0 & v(\tau) & 0 & 0 \\
0 & 0 & 0 & 1 \\
0 & 0 & -V''(\theta_0) & -v(\tau)/2
\end{pmatrix}
$$

where $V''(\theta_0) = (1-\mu)\mu^{-3/2}$ and from (7) we have that
$r(\tau) = V(\theta_0)/(h.\cosh^2((-V(\theta_0)/2)^{1/2}\tau))$, $v(\tau) = -(-2V(\theta_0))^{1/2}\tanh((-V(\theta_0)/2)^{1/2}\tau)$.
    The eigenvalues of E are: v, v, $w_{\pm} = (-v \pm (v^2 - 16V''(\theta_0))^{1/2})/4$.

It is clear that the planes $\{(r,v,\theta,u): u=0, \theta=0\}$ and $\{(r,v,\theta,u): u=0, \theta=\pi\}$ meet $\bar{I}_h$ transversally. So, the eigenvalues of E restricted to $\bar{I}_h$ are given by : v, w_+, w_-. By using E it follows that the tangent space at a point $p \in \gamma_h(\theta_0)$ splits in direct sum of a line L and an ortogonal plane π independently on the point p. The line L is generated by the eigenvector associated to the eigenvalue v and the plane π by the eigenvectors associated to the eigenvalues w_+ and w_-.

Let $_\eta *$ be the solution of the equation

$$
\eta' = \begin{pmatrix}
0 & 1 \\
-V''(\theta_0) & -v(\tau)/2
\end{pmatrix} \eta
\tag{8}
$$

with initial conditions at the unstable eigenvector of:

$$\begin{pmatrix} 0 & 1 \\ -V''(\theta_o) & -v(\tau)/2 \end{pmatrix} \tag{9}$$

when $\tau = -\infty$. By symmetry S_o transversality can be lost only when η^* turns through an angle equal to a multiple of $\pi/2$ going from $p^+(\theta_o)$ to $\gamma_h(\theta_o) \cap \{v=0\}$ (that is, from $\tau = -\infty$ to $\tau = 0$).

Now, we shall prove that the angle turned is less than $\pi/2$.

Introducing polar coordinates $\eta = (\rho.\cos\Phi, \rho.\sin\Phi)$ in (8) we obtain,

$$\Phi' = -V''(\theta_o)\cos^2\Phi - \sin^2\Phi + \beta.\tanh(\beta\tau)\sin\Phi\cos\Phi \tag{10}$$

where $\beta = (-V(\theta_o)/2)^{1/2}$. The initial conditions (9) are now given by $\tau = -\infty$ and $\Phi = \Phi_o$ where Φ_o is the unique value in $(0, \pi/4)$ such that,

$$0 = -V''(\theta_o)\cos^2\Phi_o - \sin^2\Phi_o - \beta.\sin\Phi_o\cos\Phi_o.$$

For each value of $\tau \in (-\infty, 0)$ we have an angle $\Phi^*(\tau) \in (0, \pi/4)$ such that $\Phi'(\tau, \Phi^*(\tau)) = 0$ while $\Phi'(\tau, \Phi) > 0$ if $\Phi \in (0, \Phi^*(\tau))$. From (10) it is easy to compute that $\Phi^*(\tau)$ is monotonically increasing.

Let $\Phi(\tau)$ be the solution of (10) satisfying $\Phi(-\infty) = \Phi_o$. Now, we claim that $\Phi(\tau)$ is monotonically increasing for $\tau \in (-\infty, 0)$.

By analyticity in a neighbourhood $U = (-\infty, \tau_o)$ of $\tau = -\infty$ we have either $\phi'(\tau) > 0$ or $\phi'(\tau) < 0$ for all $\tau \in U$. Suppose that $\phi'(\tau) < 0$ for all $\tau \in U$. Since $\phi(-\infty) = \phi_o = \phi^*(-\infty)$ then $0 < \phi(\tau) < \phi_o < \phi^*(\tau)$ for $\tau \in U$ and τ_o such that $\phi(\tau) > 0$ for all $\tau < \tau_o$. By definition of $\phi^*(\tau)$ we have that $\phi'(\tau) > 0$ for $\tau \in U$, and this is a contradiction. Let τ_1 be the smallest τ $(-\infty, 0)$ such that $\phi'(\tau_1) = 0$ and for all value $\tau > \tau_1$ sufficiently close to τ_1, $\phi'(\tau) < 0$.

Since $\phi(\tau) < \phi^*(\tau)$ if $\tau < \tau_1$, we have that $\phi(\tau_1) \leqslant \phi^*(\tau_1)$. This implies $\phi(\tau_1) = \phi^*(\tau_1)$. Now, there exists τ_2 in a neigbourhood of τ_1 such that $\phi'(\tau_2) < 0$. Then, $\phi(\tau_2) < \phi(\tau_1) = \phi^*(\tau_1) < \phi^*(\tau_2)$. So, $\Phi'(\tau_2) > 0$ and this is again a contradiction. Hence, $\phi(\tau)$ is monotonically increasing and $\phi(\tau) < \phi^*(\tau)$ for all $\tau \in (-\infty, 0)$.

This implies $\Phi(0) - \Phi(-\infty) < \pi/4$ as we wanted.

Q.E.D.

REMARK 2. *From the proof of Theorem 7 it follows that the angle rotated by the tangent vector to $W^\mu(p^+(0))$ on $\gamma_h(0)$ between the points $p^+(0)$ and $\gamma_h(0) \cap \{v=0\}$ is less than $\pi/4$.*

<u>THEOREM 8</u> *(i) If* $\theta_0=0,\pi$ *then* $W^\mu(p^+(\theta_0))$ *meets transversally* $W^s(P^+(\theta_0))$ *along the homothetic orbit* $\gamma_h^\infty(\theta_0)$ *on* $\overline{I}_h$ *with* $h \geqslant 0$.

(ii) If $\theta_0=0,\pi$ *then* $W^s(p^-(\theta_0))$ *meets transversally* $W^\mu(P^-(\theta_0))$ *along the homothetic orbit* $\gamma_h^\infty(\theta_0)$ *on* $\overline{I}_h$ *with* $h \geqslant 0$.

<u>Proof</u>. By using the symmetry S_0 the case (ii) follows from (i).The notation will be as in the proof of Theorem 7.

First we assume that h=0. In the variational matrix E along $\gamma_0^{0\infty}(\theta_0)$ we have:

$$r(\tau) = r(0). \exp(2^{1/2}.\mu^{-1/4}.\tau),$$
$$v(\tau) = 2^{1/2}.\mu^{-1/4}$$

Since the stable and unstable eigenvectors on the plane π are independent of the point of the homothetic orbit, (i) follows for h=0.

Now, we suppose that $h > 0$. Then in the variational matrix E along $\gamma_h^{0\infty}(\theta_0)$ we have:

$$r(\tau) = 4h^{-1}.\mu^{-1/2}.A(1-A)^{-2}, \text{ where } A=\text{constant}.\exp(2^{1/2}.\mu^{-1/4}\tau),$$
$$v(\tau) = (2(hr(\tau) + \mu^{-1/2}))^{1/2}.$$

Note that $r(\tau)$ (and hence $v(\tau)$) reach infinity for a finite value of τ, given by constant. $\exp(2^{1/2} \mu^{-1/4} \tau)=1$.

Introducing polar coordinates $\eta= (\rho.\cos\phi, \rho.\sin\phi)$ in the variational equations restricted to plane π, we obtain,

$$\phi' = -V''(\theta_0) \cos^2\phi - \sin^2\phi - (v(\tau)/2).\sin\phi.\cos\phi \qquad (11)$$

Let $\phi^u(\tau)$ (resp. $\phi^s(\tau)$) be the angle of the unstable (resp. stable) eigenvector associated to $W^u(p^+(\theta_0))$ (resp. $W^s(P^+(\theta_0))$) in the point $(r(\tau), v(\tau), \theta(\tau), u(\tau)) \in \gamma_h^{0\infty}(\theta_0)$. The functions $\phi^u(\tau)$ and $\phi^s(\tau)$ are solutions of (11).

For each value $\tau \in (-\infty, +\infty)$ we have two angles $\phi*(\tau) \in (0,\pi/4)$ and $\phi**(\tau) \in (\pi/2,2\pi)$ such that $\phi'(\tau,\phi*(\tau))=\phi'(\tau,\phi**(\tau))=0$ while $\phi'(\tau,\phi)>0$ if $\phi \in (0,\phi*(\tau)) \cup (\phi**(\tau),2\pi)$ and $\phi'(\tau,\phi)<0$ if $\phi \in (\phi*(\tau),\phi**(\tau))$. Furthermore, from (11) it is easy to see that $\phi*(\tau)$ and $\phi**(\tau)$ are monotonically decreasing.

The initial conditions of $\Phi^u(\tau)$ and $\Phi^s(\tau)$ are $\Phi^u(-\infty) = \Phi*(-\infty) \in \langle 0, \pi/4 \rangle$ and $\Phi^s(+\infty) = \Phi**(+\infty) = \pi/2$. Then using similar arguments to the proof of Theorem 7, we obtain that $\Phi^u(\tau)$ and $\Phi^s(\tau)$ are monotonocally decreasing and they satisfy, $0 < \Phi*(\tau) < \Phi^u(\tau) < \Phi*(-\infty) < \pi/2 = \Phi**(+\infty) < \Phi^s(\tau) < \Phi**(\tau) < 2\pi$. This implies that $\Phi^u(\tau) < \Phi^s(\tau)$ for all $\tau \in \langle -\infty, +\infty \rangle$, and so (i) for $h>0$.

Q.E.D.

REMARK 3. *From Theorem 7 we have that the homothetic orbits* $\gamma_h(\theta_o)$ *on* $\overline{I}_h$ *with* $h<0$ *and* $\theta_o = 0, \pi$ *are transversal. Then, by using the results due to Smale [S1], Alekseev [A1,2] and Moser [Mo] we know that some shift automorphism is a subsystem of a convenient Poincaré map defined on a surface of section transversal to* $\gamma_h(\theta_o)$ *on* $\overline{I}_h$. *In fact, Chapter IV is devoted to describe these Bernoulli's subshifts and their geometrical interpretation.*

When $h \geqslant 0$ we shall see in Chapter III that all solutions escape to infinity. Then there are no recurrent orbits and it is not possible to put the shift automorphism as a subsystem of a Poincaré map associated to the transversal homothetic orbits $\gamma_h^{o\infty}(\theta_o)$ and $\gamma_h^{\infty o}(\theta_o)$ for $\theta_o = 0, \pi$ studied in Theorem 8.

(II.7) The flow on the collision manifold.

From Proposition 2 all the equilibrium points are hyperbolic for the flow (2). Thus the Hartmann-Grobman theorem, describes the local behaviour of the flow at these points. From now on, in this section, we restrict the flow (2) on Λ. Hence, we have that:

$p^+_-(0)$, $p^+_-(\pi)$ are saddles,
$p^-(\pi/2)$, $p^-(-\pi/2)$ are sources if $\mu \in (1, 9/8]$ and unstable foci if $\mu > 9/8$
$p^+(\pi/2)$, $p^+(-\pi/2)$ are sinks if $\mu \in (1, 9/8]$ and stable foci if $\mu > 9/8$.

If we compute the symmetries given in (II.2) in coordinates (r, v, θ, u, τ) we obtain:

$$S_o(r, v, \theta, u, \tau) = (r, -v, \theta, -u, -\tau),$$
$$S_1(r, v, \theta, u, \tau) = (r, -v, -\theta, u, -\tau) \text{ and}$$
$$S_2(r, v, \theta, u, \tau) = (r, -v, \pi-\theta, u, -\tau).$$

Of course, every composition of these symmetries give us another symmetry which leaves invariant the flow (2). For example, the symmetry $S_3 = S_2 \circ S_1$ given by,

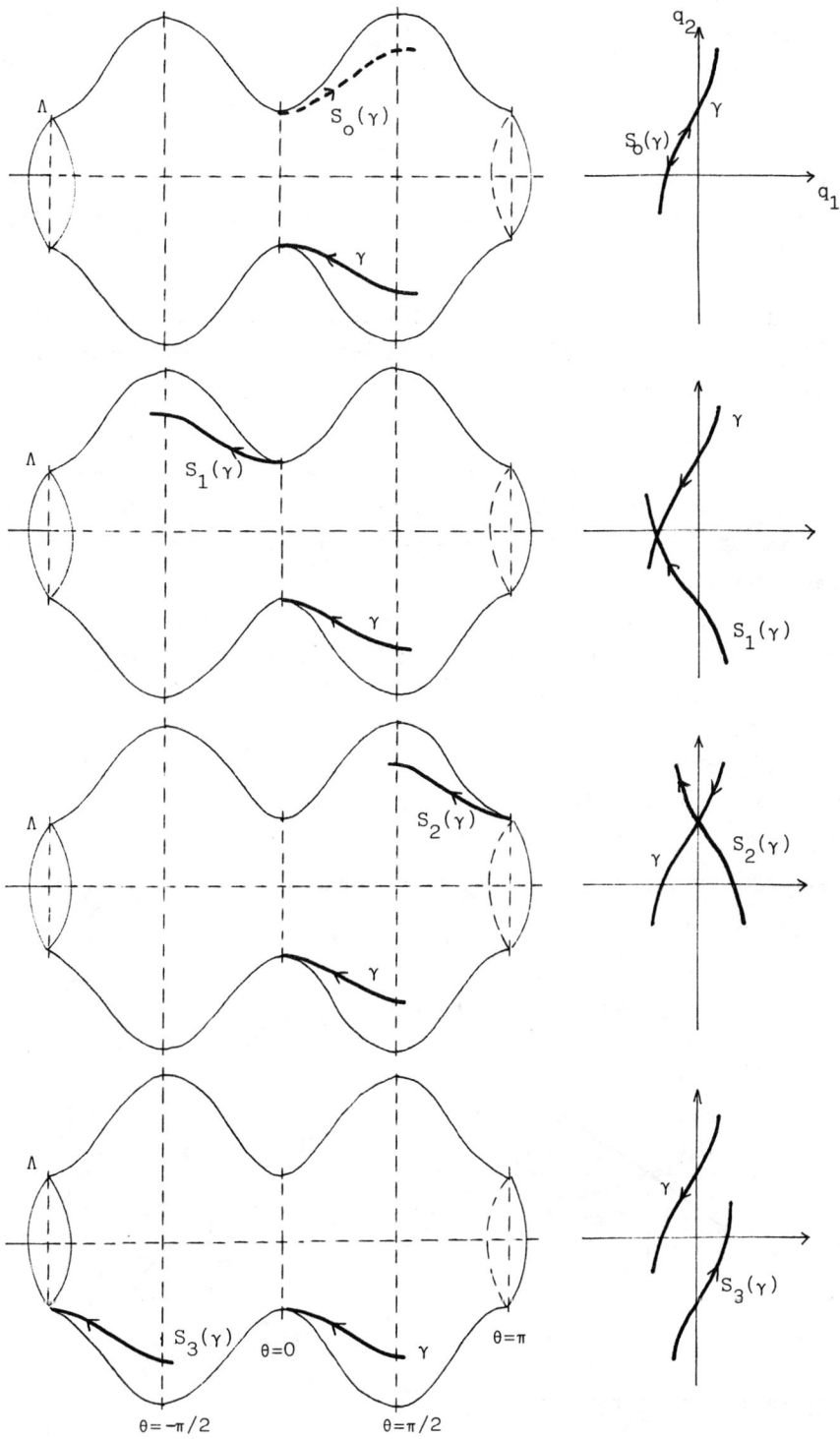

Figure 8. Geometrical interpretation of the symmetries S_i, i=0,1,2,3, on the collision manifold Λ and on the configuration space. On Λ the continuous (resp discontinuous) curve is used when the solution has u>0 (resp. u<0).

$$S_3(r,v,\theta,u,\tau) = (r,v,\pi+\theta,u,\tau)$$

leaves invariant the flow (2) but does not satisfy the condition (2) of reversing involution. For a geometric interpretation of these symmetries, see Figure 8.

In order to study the local flow at the four equilibrium points $p^+_-(0)$,
$p^+_-(\pi)$ (resp. $p^+_-(\pi/2)$, $p^+_-(-\pi/2)$) it is sufficient to study one of them and to
use the symmetries.

For a saddle point p we denote its four invariant branches in the following
way : $B^{u,s}_{+,-}(p,\mu)$ will be the unstable (u) or stable (s) invariant branch contained in u>0 (+) or in u<0 (-) in a neigbourhood of p and for the value μ of the
parameter.

If we compute the eigenvalues λ^u, λ^s and their eigenvectors ω^u, ω^s at the
critical point $p^-(0)$, then we obtain:

$$\lambda^u = 2^{-3/2} \mu^{-1/4}(1+(9-8\mu^{-1})^{1/2}), \quad \omega^u =(1, \lambda^u) \quad \text{and}$$
$$\lambda^s = 2^{-3/2} \mu^{-1/4}(1-(9-8\mu^{-1})^{1/2}), \quad \omega^s =(1, \lambda^s).$$

Therefore, Figure 9 give us the local behaviour of the four invariant branches
$B^{u,s}_{+,-}(p^-(0),\mu)$.

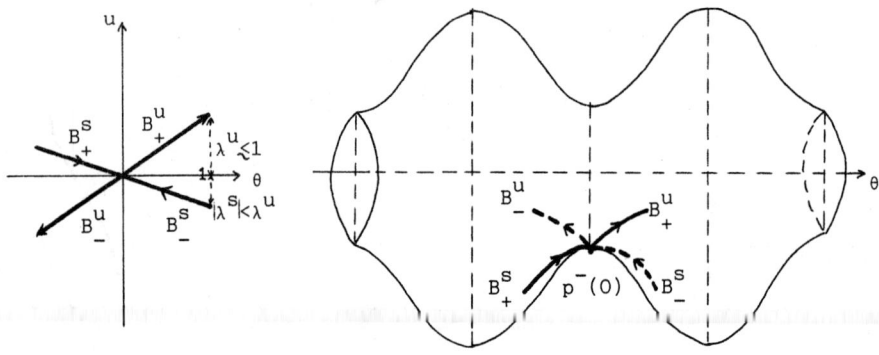

Figure 9. Local behaviour of $B^{u,s}_{+,-}(p^-(0),\mu)$.

Recall that a vector field is called <u>gradient</u> <u>like</u> with respect to a function v if v increases along all non-equilibrium point orbits. From (2) and (3) we have that the flow on Λ satisfies $v'=u^2/2$ and Lemma 9 follows,

<u>LEMMA</u> 9 (Devaney [D2,5]). *The flow on Λ is gradient like with respect to the v-coordinate.*

Let p be a saddle point of Λ . Set $P_{+,-}^{u,s}(p,\mu)= B_{+,-}^{u,s}(p,\mu) \cap \{v=0\}$. Note that from Lemma 9 $P_{+,-}^{u,s}(p,\mu)$ is a unique point.

In order to describe the global qualitative behaviour of the flow on Λ it is sufficient to study the sixteen branches $B_{+,-}^{u,s}(p,\mu)$. The knowledge of the sixteen points $P_{+,-}^{u,s}(p,\mu)$ will be enough.

<u>LEMMA</u> 10. *For all $\mu\in(1,\infty)$ we have that $\pi/2 \leqslant \theta(P_+^u(p^-(0),\mu))< \pi$.*

<u>Proof</u>. From (2) and (3) we obtain that $d\theta/dv=2(-2V(\theta)-v^2)^{-1/2}$ on $\Lambda\cap\{u\geqslant 0\}$. Since $\min\limits_{\theta} (-V(\theta)) = \mu^{-1/2}$, we have that,

$$\Delta\theta = \int_{-2^{1/2}\mu^{-1/4}}^{0} 2(-2V(\theta)-v^2)^{-1/2} \, dv \;<\; \int_{-2^{1/2}\mu^{-1/4}}^{Q} 2(2\mu^{-1/2}-v^2)^{-1/2} \, dv \;=\; \pi.$$

Then for all $\mu\in(1,+\infty)$, $\theta(P_+^u(p^-(0),\mu)) < \pi$.

Now, we assume that $\theta(P_+^u(p^-(0), \mu_0)) < \pi/2$ for some $\mu_0 \in (1,+\infty)$. By using symmetry S_2 and Lemma 9) we obtain the behaviour of $B_+^u(p^-(0), \mu_0)$, $B_-^s(p^-(0), \mu_0)$, $B_-^u(p^+(\pi), \mu_0)$ and $B_+^s(p^+(\pi), \mu_0)$ as in Figure 10.

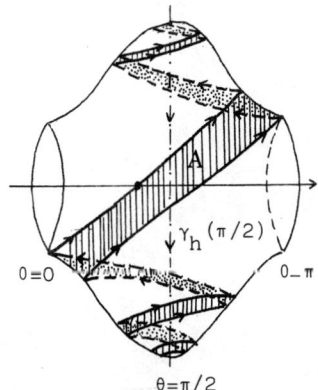

$\theta=\pi/2$

<u>Figure 10</u>. Behaviour of $B_+^u(p^-(0), \mu_0)$, $B_-^s(p^-(0), \mu_0)$, $B_-^u(p^+(\pi), \mu_0)$ and $B_+^s(p^+(\pi), \mu_0)$ when $\theta(P_+^u(p^-(0), \mu_0))< \pi/2$.

Since the flow on Λ gives us the behaviour of the solutions close to $r=0$, we have that there are orbits which go near the homothetic orbit $\gamma_h(\pi/2)$ and after they follow closely to the flow on the region A of Figure 10. For such an orbit γ its projection on the configuration plane looks like Figure 11. This is a contradiction because our problem is a central force problem and γ does not cross the q_1-axis.

Q.E.D.

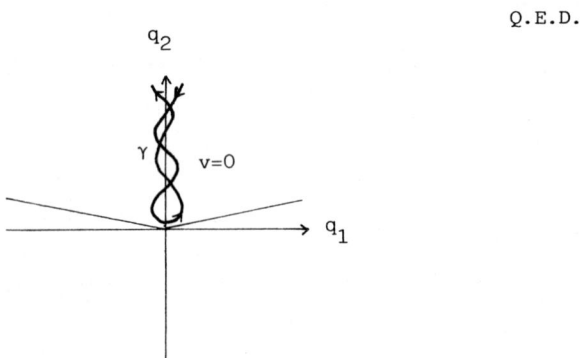

Figure 11.The projection of the orbit γ near $\gamma_h(\pi/2)$ and the region A of
          Figure 10.

In fact, numerical computations of the function $\theta(\mu) = \theta(P_+^u(p^-(0),\mu))$ show that $\pi/2 < \theta(\mu) < \pi$ and decreases in $(1,\infty)$, see Figure 12.

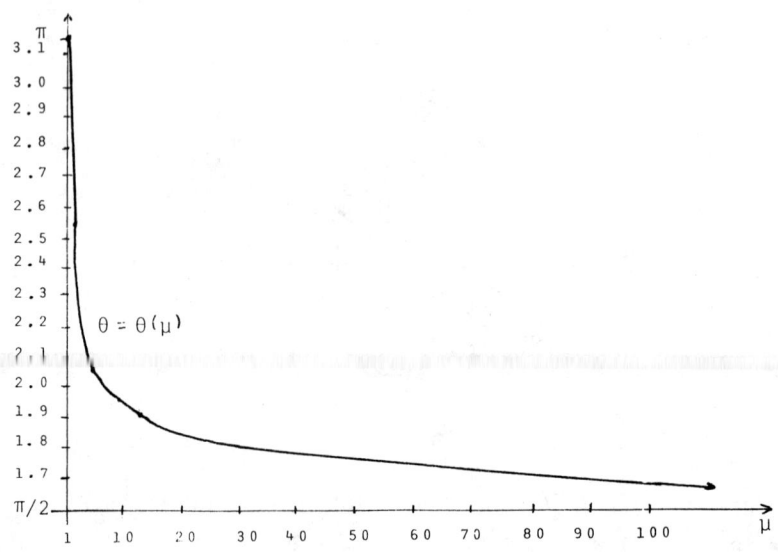

Figure 12. The function $\theta(\mu)=\theta(P_+^u(p^-(0),\mu))$.

<u>LEMMA 11</u>. *For all* $\mu > 1$, *Figure 13 describes the position of the eight points* $P_+^{u}(p^-(0),\mu)$, $P_-^{u}(p^-(0),\mu)$, $P_+^{u}(p^-(\pi),\mu)$, $P_-^{u}(p^-(\pi),\mu)$, $P_+^{s}(p^+(0),\mu)$, $P_-^{s}(p^+(0),\mu)$, $P_+^{s}(p^+(\pi),\mu)$ *and* $P_-^{s}(p^+(\pi),\mu)$.

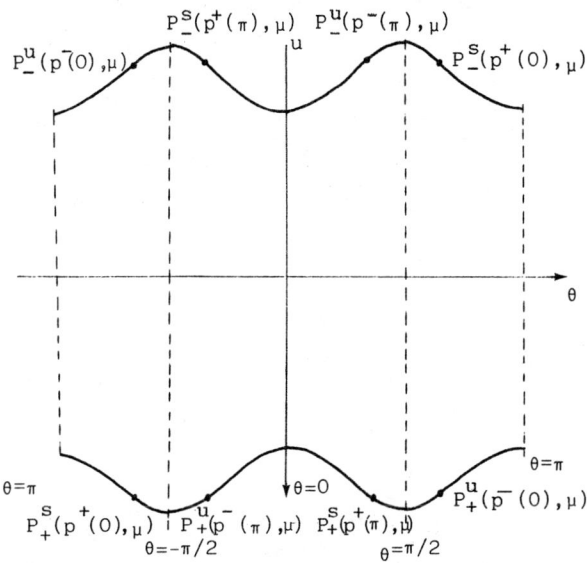

<u>Figure 13</u>. The points $P_{+,-}^{u,s}(p_-^+(\theta_0),\mu)$ where $\theta_0 = 0,\pi$.

The following theorem improves Theorem 2.7 of [D3] and Theorem 4.10 of [D2] .

<u>THEOREM 12</u>. *(i) For all* $\mu \in (1,\infty)$ *we have that the orbits* $B_+^{u}(p^+(0),\mu)$, $B_-^{u}(p^+(0),\mu)$, $B_+^{u}(p^+(\pi),\mu)$ *and* $B_-^{u}(p^+(\pi),\mu)$ *are forward asymptotic to* $p^+(\pi/2)$, $p^+(-\pi/2)$, $p^+(-\pi/2)$, *and* $p^+(\pi/2)$ *respectively, and the orbits* $B_+^{s}(p^-(0),\mu)$, $B_-^{s}(p^-(0),\mu)$, $B_+^{s}(p^-(\pi),\mu)$ *and* $B_-^{s}(p^-(\pi),\mu)$ *are backward asymptotic to* $p^-(-\pi/2)$, $p^-(\pi/2)$, $p^-(\pi/2)$ *and* $p^-(-\pi/2)$ *respectively. See Figures 14.*

(ii) For all $\mu \in (1,\infty)$ *we have that the orbits* $B_+^{s}(p^+(0),\mu)$, $B_-^{s}(p^+(0),\mu)$, $B_+^{s}(p^+(\pi),\mu)$ *and* $B_-^{s}(p^+(\pi),\mu)$ *are backward asymptotic to* $p^-(\pi/2)$, $p^-(-\pi/2)$, $p^-(-\pi/2)$ *and* $p^-(\pi/2)$ *respectively, and the orbits* $B_+^{u}(p^-(0),\mu)$, $B_-^{u}(p^-(0),\mu)$, $B_+^{u}(p^-(\pi),\mu)$ *and* $B_-^{u}(p^-(\pi),\mu)$ *are forward asymptotic to* $p^+(-\pi/2)$, $p^+(\pi/2)$, $p^+(\pi/2)$ *and* $p^+(-\pi/2)$ *respectively. See Figures 14*

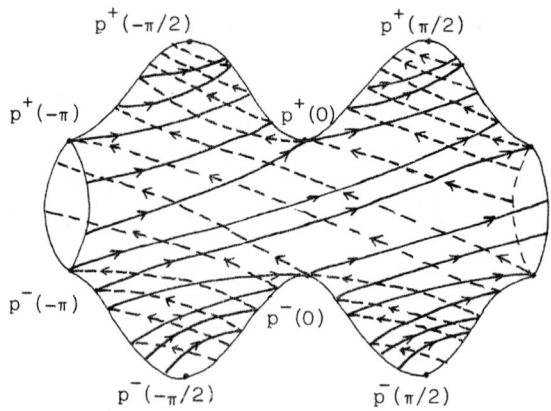

<u>Figure 14a</u>. The flow on Λ for μ>9/8.

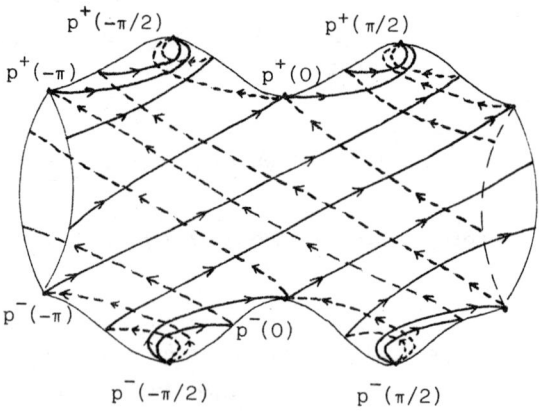

<u>Figure 14b</u>. The flow on Λ for 1<μ⩽9/8.

<u>Proof</u>. Part (i) follows from the fact that the flow on Λ is gradient like with respect to the v-coordinate.

Part (ii) follows from Lemmas 9,10,11 and the symmetries S_i.

Q.E.D.

Note that Theorem 12 gives us the global flow on Λ for all μ∈(1,∞).

The following corollary extends the results of Devaney [D3].

COROLLARY 13. For $\mu \in (1,\infty)$ the anisotropic Kepler problem cannot be regularized (in the sense of Easton [E]).

Proof. From Theorems 7 and 8 and Figure 14, the solutions in a neigbourhood of the homothetic orbits $\gamma_h(\theta_o)$ and $\gamma_h^{\infty o}(\theta_o)$ for $\theta_o = 0, \pi$, cannot be extended in any continuous fashion after passing near collision.

Q.E.D.

Now, we shall study the number of revolutions of $B_{\pm}^{u}(p^+(\theta_o),\mu)$ or $B_{\pm}^{s}(p^-(\theta_o),\mu)$ for $\theta_o = 0, \pi$ around the equilibrium points to which they tend when $\tau \to +\infty$ or $\tau \to -\infty$ (see Theorem 12), respectively. By using the symmetries S_i is sufficient to study this behaviour for the branch $B_+^{s}(p^-(\pi),\mu)$ when it tends to $p^-(\pi/2)$ for $\tau \to -\infty$.

In order to compute the revolutions around the point $p^-(\pi/2)$ we introduce polar coordinates centered in this point. That is, $u = \rho.\sin\phi$, $\theta - \pi/2 = \rho.\cos\phi$.

Let $p_\mu(\tau) = (v_\mu(\tau), \rho_\mu(\tau), \Phi_\mu(\tau)) = (v_\mu(\tau), \theta_\mu(\tau), u_\mu(\tau))$ be the solution corresponding to the branch $B_+^{s}(p^-(\pi),\mu)$ such that $p_\mu(\tau) \to p^-(\pi)$ when $\tau \to +\infty$ and $p_\mu(\tau) \to p^-(\pi/2)$ when $\tau \to -\infty$. Then the <u>number of revolutions</u> $R(\mu)$ of the branch $B_+^{s}(p^-(\pi),\mu)$ around $p^-(\pi/2)$ is define by:

$$R(\mu) = (\lim_{\tau \to +\infty} \Phi_\mu(\tau) - \lim_{\tau \to -\infty} \Phi_\mu(\tau))/2\pi \ .$$

PROPOSITION 14. (i) $R(\mu) = +\infty$ if $\mu \in (9/8, +\infty)$
(ii) $R(\mu) = 0$ if $\mu \in (1, 9/8]$

Proof. By Corollary 3, (i) follows. Since $p^-(\pi,\mu)$ is a saddle and $p^-(\pi/2,\mu)$ is a source without spiralling when $\mu \in (1, 9/8]$, (ii) follows from Figure I.1 and the local behaviour of the flow near $p^-(\pi/2)$ (see (V.1) and Figure V.1 for more details), see Figure 14b.

Q.E.D.

III. THE FLOW FOR NON-NEGATIVE ENERGY LEVELS

(III.1) The case h=0

The invariant manifold $\bar{I}_o$ (see (II.5) and Figure II.7b) is formed by the manifold I_o and the boundary submanifolds Λ and N_o. The equations of motion in $I_o \cup \Lambda$ are,

$$
\begin{aligned}
r' &= rv \\
v' &= u^2/2 \\
\theta' &= u \\
u' &= -vu/2 - V'(\theta)
\end{aligned}
\tag{1}
$$

and in $I_o \cup N_o$ the same equations except the first one which becomes $\rho' = -\rho v$.

System (1) can be solved in the variables (v, θ, u) and after in the variable r or ρ. This means that the flow on I_o is projectable on Λ or N_o. Since we know the flow on Λ or N_o (see Theorem II.12 and Figure II.14), the solutions on I_o can be obtained lifting the solutions on Λ or N_o in the radial direction r or ρ, respectively.

The homothetic solutions $\gamma_o(\theta_o)$ for $\theta_o = 0,\ \pi/2,\ \pi,\ -\pi/2$ connect the boundaries Λ and N_o, and they are the unique solutions on I_o whose projection on Λ or N_o is a point (see Figure II.7).

Let $W^s(\Lambda)$ (resp. $W^u(\Lambda)$) denote the set of points in I_o whose forward (resp. backward) orbits converge to Λ. Similarly, we define $W^s(N_o)$ and $W^u(N_o)$.

PROPOSITION 1. (i) $W^s(N_o) = I_o \setminus W^s(\Lambda)$ and $W^u(N_o) = I_o \setminus W^u(\Lambda)$.
(ii) $W^s(\Lambda) \subset W^u(N_o)$ and $W^u(\Lambda) \subset W^s(N_o)$; see Figure 1.

Proof. Note that $W^s(\Lambda)$ is formed by all the collision orbits and $W^u(\Lambda)$ by all the ejection orbits. On the other hand, Table II.2 give us $\dim(W^s(N_o)) = 3 = \dim(W^u(N_o))$. Then Proposition 1 follows.

$$Q.E.D.$$

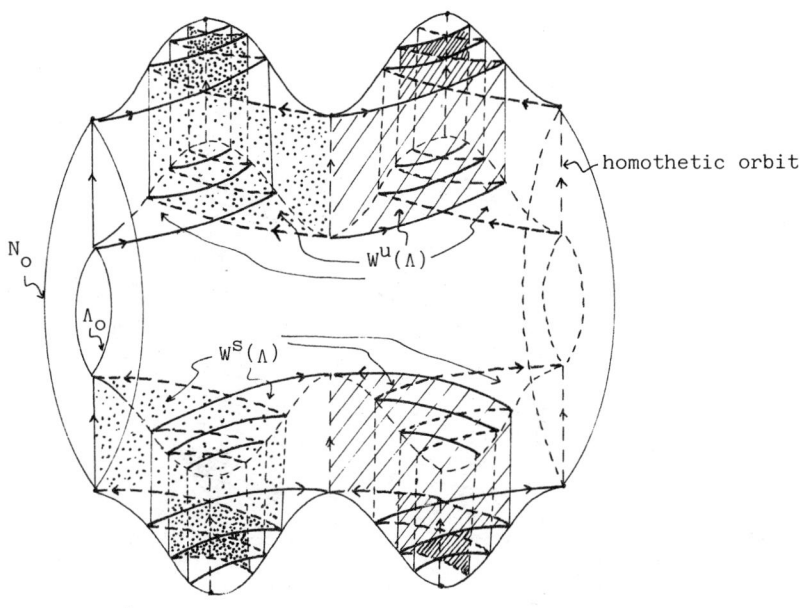

Figure 1. The sets $W^S(\Lambda) = W^S_{+,-}(p^-(0)) \cup W^S_{+,-}(p^-(\pi))$ and
$W^u(\Lambda) = W^u_{+,-}(p^+(0)) \cup W^u_{+,-}(p^+(\pi))$, for $\mu > 9/8$.

THEOREM 2. *(i) On the manifold Λ_o or N_o we denote by $A, B, C,\ D$ (resp. $A', B', C',\ D'$) the closed strips shown in Figure 2 (resp, Figure 3). Each strip corresponds on I_o to a different qualitative behaviour of the solutions, whose projections on the configuration plane are given in Figures 4.*

(ii) The collision and ejection orbits are projected on the branches $C' \cap B$, $D' \cap A$, $A' \cap D$, $B' \cap C$ and $D' \cap C$, $A' \cap B$, $B' \cap A$, $C' \cap D$ respectively, i.e., on the branches of the stable and unstable manifolds of equilibrium points $p^{+,-}(\theta_o)$ with $\theta_o = 0, \pi/2,\ \pi, -\pi/2$. Their geometrical behaviour on the configuration plane is described in Figures 5.

(iii) If $\mu > 1$ then the escape (resp. capture) solutions tend to infinity when $\tau \to +\infty$ (resp. $\tau \to -\infty$) in the directions $\theta_o = 0, \pi/2,\ \pi,\ -\pi/2$ with radial velocity $v = (-2V(\theta_o))^{1/2}$ (resp. $v = -(-2V(\theta_o))^{1/2}$).

(iv) In Figures 4 and 5 the number of oscillations around the q_2-axis is zero or infinite depending on whether $\mu \in (1,\ 9/8]$ or $\mu \in (9/8, \infty)$.

Proof. Part (i) and (ii) follow from Theorem II.12 and Figure II.14. By Proposition II. 4 and II.14 we obtain parts (iii) and (iv), respectively.

Q.E.D.

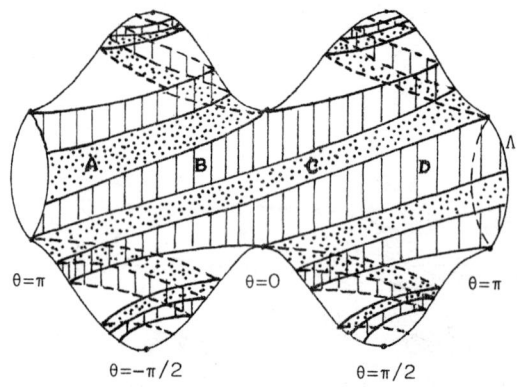

Figure 2. The regions A,B,C and D on Λ or N_o for $\mu > 9/8$.

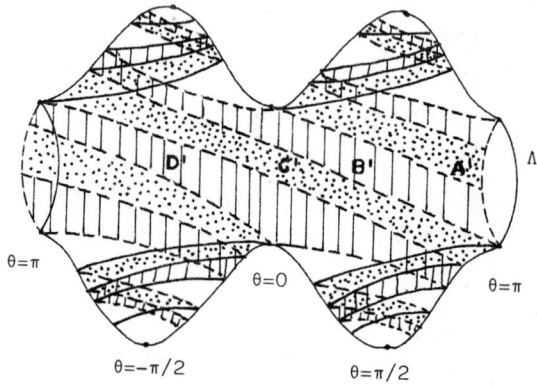

Figure 3. The regions A',B',C' and D' on Λ or N_o for $\mu > 9/8$.

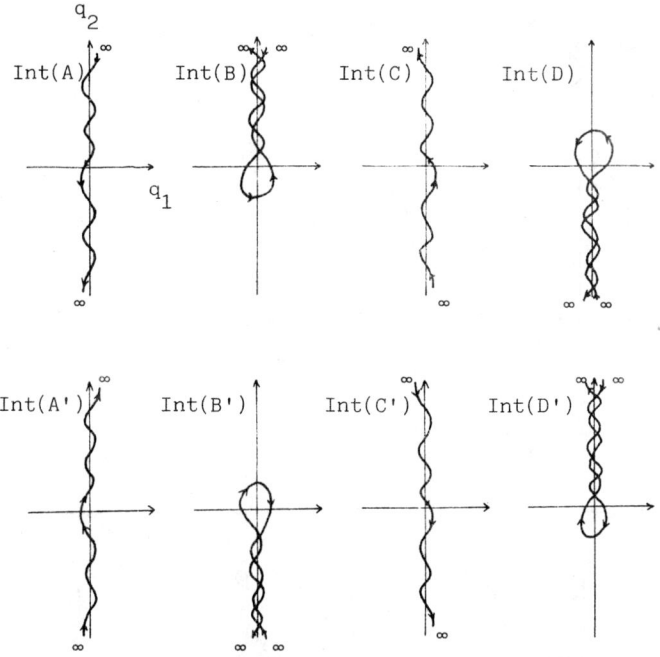

Figure 4a. Qualitative behaviour of the orbits of I_o whose projection on Λ or N_o lies on A,B,C,D,A',B',C' or D' for $\mu > 9/8$. Here Int(X) means the interior of the set X.

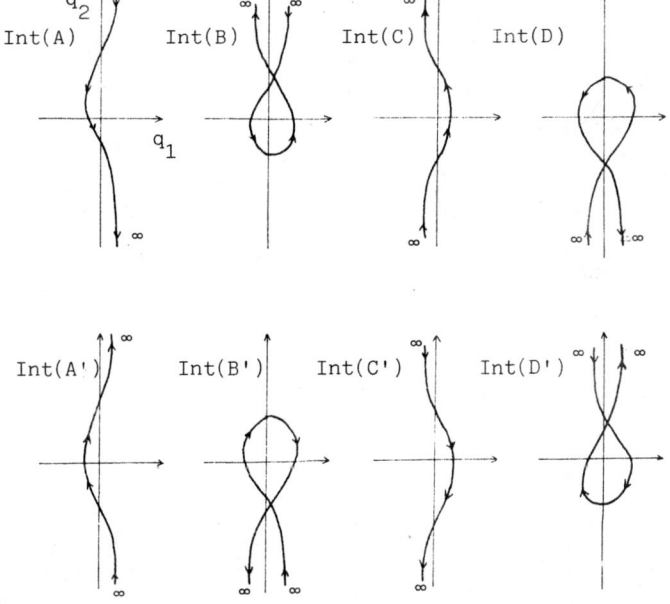

Figure 4b. Qualitative behaviour of the orbits of I_o whose projection on Λ or N_o lies on A,B,C,D,A',B',C' or D' for $1 < \mu \leqslant 9/8$.

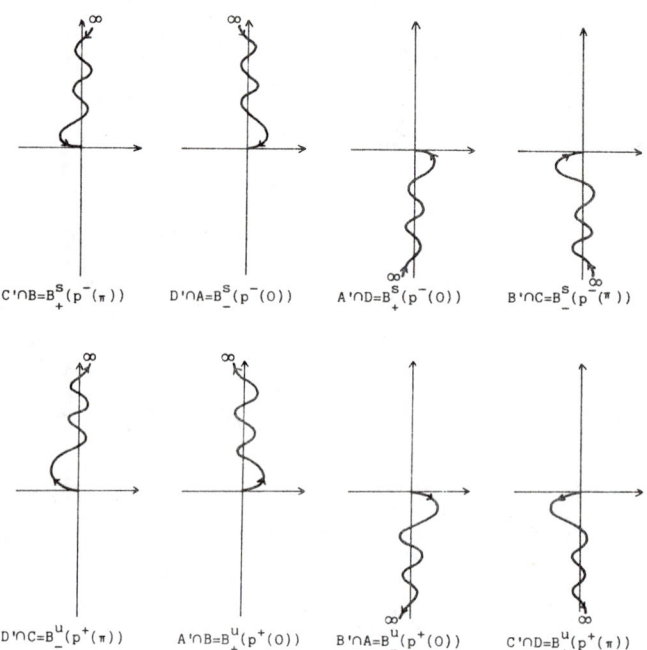

Figure 5a. Qualitative behaviour of the orbits of I_o whose projection on Λ or N_o lies on $B_{+,-}^{u,s}(p^{+,-}(\theta_o))$ with $\theta_o = 0$ or π, for $\mu > 9/8$.

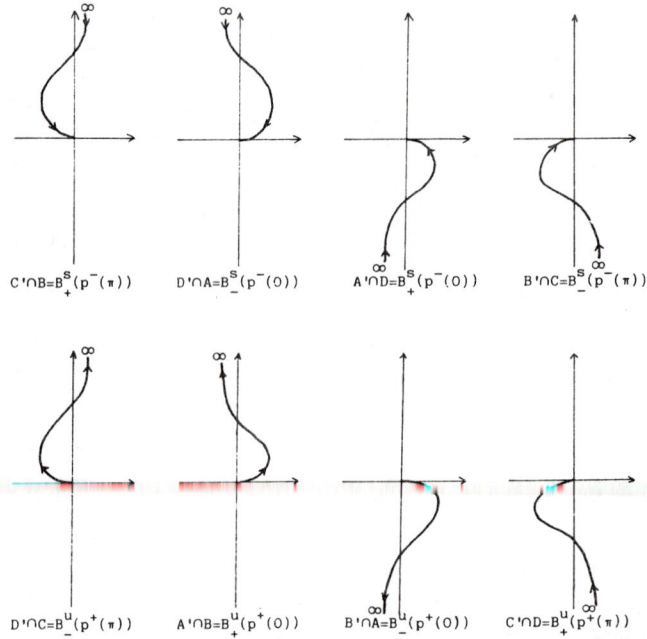

Figure 5b. Qualitative behaviour of the orbits of I_o whose projection on Λ or N_o lies on $B_{+,-}^{u,s}(p^{+,-}(\theta))$ with $\theta_o = 0$ or π, for $\mu \in (1, 9/8]$.

(III.2) The case h>0.

Similarly to the case h=0, we have that $\overline{I}_h$ (see (II.5) and Figure II.7) is formed by the manifold I_h and the boundary submanifolds Λ and N_h. Now, the equations of motion in $I_h \cup \Lambda$ are,

$$r' = rv$$
$$v' = u^2/2 + rh$$
$$\theta' = u$$
$$u' = -vu/2 - V'(\theta)$$

and in $I_h \cup N_h$ are (see (II.4)),

$$\rho' = -\rho W$$
$$W' = U^2 + \rho V(\theta)$$
$$\theta' = U$$
$$U' = -WU - \rho V'(\theta)$$

When h>0 the flow is not projectable on Λ or N_h but, in a similar way to (III.1), we can prove:

PROPOSITION 3. (i) $W^s(N_h) = I_h \setminus W^s(\Lambda)$ and $W^\mu(N_h) = I_h \setminus W^\mu(\Lambda)$.
(ii) $W^s(\Lambda) \subset W^\mu(N_h)$ and $W^\mu(\Lambda) \subset W^s(N_h)$.
(iii) If $\mu > 1$ then the escape(resp. capture) solutions have no restriction in the θ-direction when they tend to infinity, while the radial velocity satisfies $|W| = (2h)^{1/2}$.

IV. THE FLOW ON NEGATIVE ENERGY LEVELS WHEN $\mu > 9/8$.

(IV.1) The intersection of the invariant manifolds with the surface of section $v = 0$

As we said in Remark 3 of (II.6) we are interested in describing the shift automorphism as a subsystem of a Poincaré map defined on a surface of section transversal to the homothetic orbits $\gamma_h(\theta_o)$ on $\bar{I}_h$ for $\theta_o = 0$, $\pi/2$, π, $-\pi/2$ and $h<0$. This surface of section will be the annulus $S = \{(r,v,\theta,u) : v=0, (u^2+v^2)/2 + V(\theta) = rh\} = \bar{I}_h \cap \{v=0\}$ with $h<0$.

The intersection of the annulus S with the collision manifold Λ is given by

$$\Lambda \cap S = \{(r,v,\theta,u) : r=0, v=0, u^2/2 = -V(\theta)\} \text{ , see Figure 1.}$$

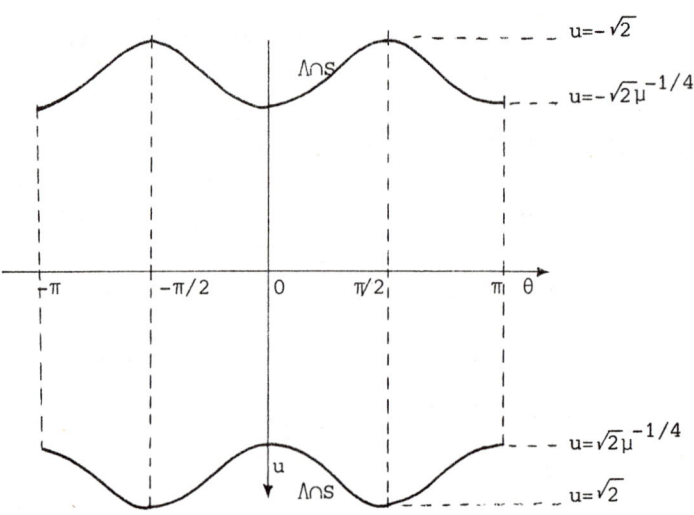

Figure 1. The set $\Lambda \cap S$.

The flow is transversal to the annulus S when $v' \neq 0$. The curves $w = \{v'=0\} \cap S = \{(\theta,u): u^2 = -V(\theta)\}$ are shown in Figure 2. The regions $\{v'>0\} \cap S$ and $\{v'<0\} \cap S$ correspond to the orbits which cross the annulus S with v increasing and v decreasing, respectively (see Figure 2 again).

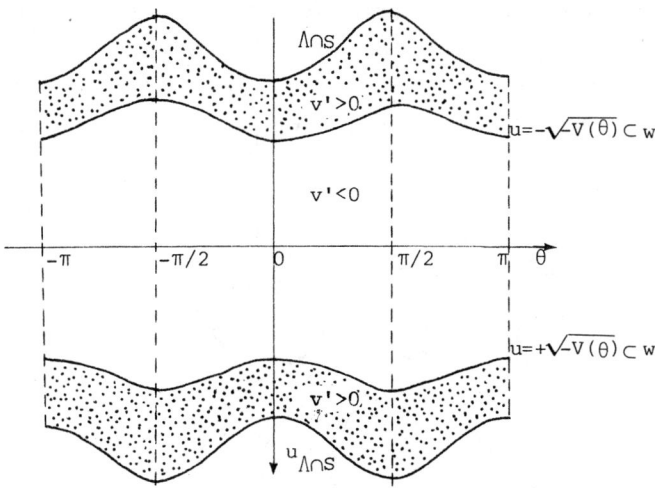

Figure 2. The curves w and the regions $\{v'>0\}\cap S$ and $\{v'<0\}\cap S$

We denote by $W^u_{+,-}(p,\mu)$ (resp. $W^s_{+,-}(p,\mu)$) the unstable (resp. stable) invariant manifold associated to the saddle point p such that $W^u_{+,-}(p,\mu) \cap \Lambda = B^u_{+,-}(p,\mu)$ (resp. $W^s_{+,-}(p,\mu) \cap \Lambda = B^s_{+,-}(p,\mu)$). Here, we use the notation introduced in (II.7).

LEMMA 1. For $\mu > 1$ the following equalities hold.

$$W^u_+(p^+(0),\mu) = S_3(W^u_+(p^+(\pi),\mu)),$$

$$W^u_-(p^+(\pi),\mu) = S_2 \circ S_0(W^u_+(p^+(0),\mu)),$$

$$W^u_-(p^+(0),\mu) = S_2 \circ S_0(W^u_+(p^+(\pi),\mu)),$$

$$W^s_+(p^-(\pi),\mu) = S_2(W^u_+(p^+(0),\mu)),$$

$$W^s_+(p^-(0),\mu) = S_2(W^u_+(p^+(\pi),\mu)),$$

$$W^s_-(p^-(\pi),\mu) = S_2(W^u_-(p^+(0),\mu)) \; and$$

$$W^s_-(p^-(0),\mu) = S_2(W^u_-(p^+(\pi),\mu))$$

The proof follows easily from the symmetries.

We define the curve $\sigma^u_{+,-}(p,\mu)$ (resp. $\sigma^s_{+,-}(p,\mu)$) as the first intersection of $W^u_{+,-}(p,\mu)$ (resp. $W^s_{+,-}(p,\mu)$) with S in forward time (resp. backward time) where p is one of the four saddle points on Λ.

From Corollary II.3 we can prove (see Proposition 5.3 of [D2]) that if $\mu > 9/8$ then there is a neighbourhood of $\gamma_h(\theta_o) \cap S$ where $\sigma_{+,-}^{u,s}(p(\theta_1),\mu)$ spirals tending to $\gamma_h(\theta_o) \cap S$, where $\theta_o = \pm \pi/2$ and $\theta_1 = 0,\pi$. Therefore in a neighbourhood of $\gamma_h(\theta_o) \cap S$ in S the curves $\sigma_{+,-}^{u,s}(p(\theta_1),\mu)$ look like those in Figure 3.

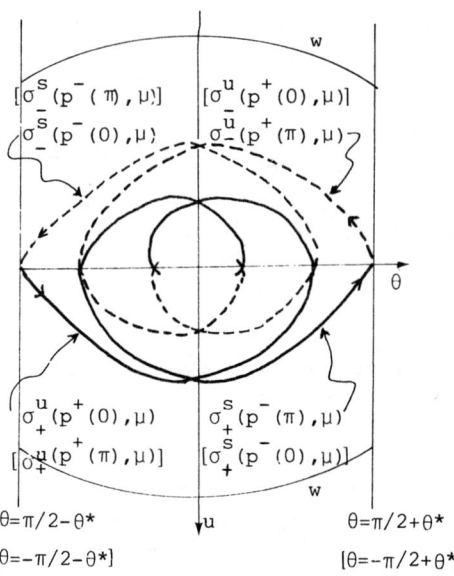

$$\theta = \pi/2 - \theta* \qquad\qquad \theta = \pi/2 + \theta*$$
$$[\theta = -\pi/2 - \theta*] \qquad\qquad [\theta = -\pi/2 + \theta*]$$

Figure 3. The curves $\sigma_{+,-}^{u,s}(p(\theta_1),\mu)$ where $\theta_1 = 0$ or π. Here $\theta* = \theta*(\mu)$ is a value of θ close enough to 0.

   From now on, we use $[\ ]$ in order to represent another case in the same picture.

   Numerical computations for some values of μ show that the curves σ^s and σ^u of Figure 3 meet transversaly. So, the analyticity of these curves proves that they meet transversaly for almost every $\mu \in (9/8,\infty)$. For our topological study of the flow (see Section IV.7) this transversality does not play any role.

   We choose an arc $\bar{\sigma}_+^u(s)$ with $s \in [0,+\infty]$ on $W_+^u(p^+(0),\mu)$ as in Figure 4, such that $\bar{\sigma}_+^u(0) \in \gamma(0)$ and $\bar{\sigma}_+^u(+\infty) \in R_+^u(p^+(0),\mu)$. This arc gives a natural parametrization $\sigma_+^u(s)$ of $\sigma_+^u(p^+(0),\mu)$ with parameter $s \in [0,+\infty)$ such that $\sigma_+^u(0) = (0,0)$ and $\lim\limits_{s \to \infty} \sigma_+^u(s) = (\pi/2, 0)$. We shall prove that $\sigma_+^u(s)$ is a continuous arc for all $s \in [0,\infty)$ when $\mu \in (9/8,4]$. Let $\gamma_s(\tau) = (r_s(\tau), v_s(\tau), \theta_s(\tau), u_s(\tau))$ be the orbit such that $\gamma_s(0) = \sigma_+^u(s)$.

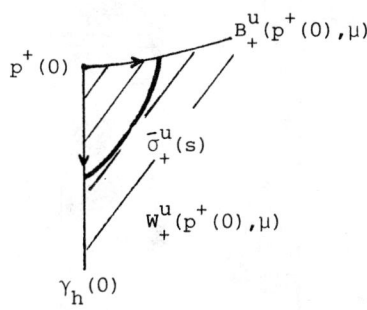

Figure 4. The arc $\bar{\sigma}_+^u(s)$.

LEMMA 2. (i) $\sigma_+^u(s)$ _in a neighbourhood of (0,0) is contained in_ $\{\theta > 0\} \cap \{u > 0\} \cap S$,
see Figure 5.

(ii) Let s* be the smallest value of s>0 such that $\sigma_+^u(s*)$ _belongs to w or_
$\{u=0\}$. _Then,_ $u_s(\tau) > 0$ _for all_ $\tau < 0$ _and for all_ $s \in (0,s*)$.

Proof. (i): By Remark II.2 and Figure II.9, for s small enough $\sigma_+^u(s)$ is an arc
contained in $\{\theta > 0\} \cap \{u > 0\} \cap S$. Then (i) follows.

(ii): The orbits $\gamma_s(\tau)$ with s>0 small enough are such that $u_s(\tau) > 0$ for all $\tau < 0$,
because they are close to $\gamma_h(0)$ and they lie on $W_+^u(p^+(0),\mu)$, see Remark II.2.

Suppose that there exists $s_1 \in (0,s*)$ such that $u_{s_1}(\tau) < 0$ for some $\tau < 0$. Let
s_2 be the smallest value of s>0 such that $u_{s_2}(\tau_2) = 0$ for some $\tau_2 < 0$. So,
$u'_{s_2}(\tau_2) = 0$. Therefore, from II-(2) we have that $\theta_{s_2}(\tau_2) \in \{0, \pi/2, \pi, -\pi/2\}$.
This implies that $\gamma_{s_2}(\tau_2) \in \gamma_h(\theta_{s_2}(\tau_2))$ and this is a contradiction. Hence
(ii) follows.

Q.E.D.

The global behaviour of $\sigma_{+,-}^{u,s}(p(\theta_1),\mu)$ for $\mu \in [9/8, \mu_c)$ for some
$\mu_c = \mu_{critical} > 4$ (see later) is given by the following theorem. In the proof we
shall use ideas given in [LMS].

THEOREM 3. For $\mu \in [9/8, 4]$ _we have that_ $\sigma_+^u(s)$ _is a continuous arc for all_
$s \in [0,\infty)$ _contained in_ $\{v' < 0\} \cap S$. _Furthermore,_ $\theta_s(0) \in [0,\pi)$ _for all_ $s \in [0,\infty)$.
See Figure 5.

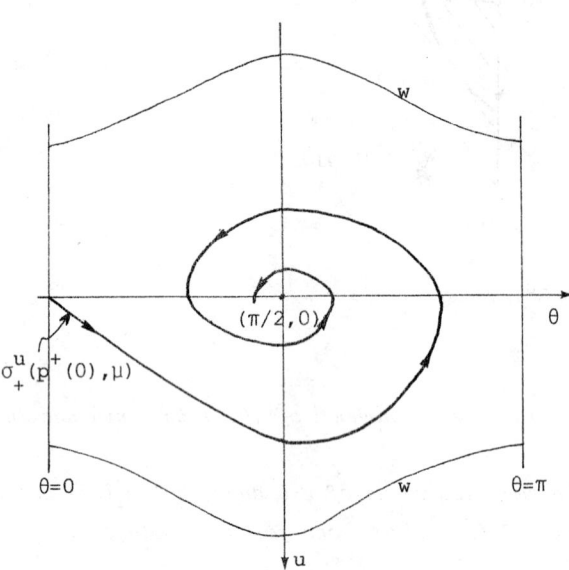

Figure 5. The curve $\sigma_+^u(p^+(0),\mu)$ for $\mu\in[9/8,4]$.

<u>Proof</u>. The continuity of the arc $\sigma_+^u(s)$ can only be lost if it meets the curve
w. First of all, suppose that the arc $\sigma_+^u(s)$ meets the curve w before crossing
the circle $\{u=0\}$. Let s* be the smallest value of s>0 such that $\sigma_+^u(s*)=(\theta*,u*)$
belongs to w. We claim that $V(\theta*) < 2V(0)$.

To prove the claim we consider the orbit $\gamma_{s*}(\tau)$. From II.(2) we have that
$du/d\theta = -v/2 -V'(\theta)/u$. Since $v_{s*}(\tau)>0$ and $u_{s*}(\tau)>0$ (by Lemma 2) for all $\tau<0$,
we obtain that $udu <-V'(\theta) d\theta$. By integration from $p^+(0)$ to $\sigma_+^u(s*)$ we obtain,

$$u^2/2 \Big]_{u=0}^{u=(-V(\theta*))^{1/2}} < -V(\theta) \Big]_{\theta=0}^{\theta=\theta*}$$. So, $V(\theta*) <2V(0)$ and we have proved the

claim.

Now, we shall prove that $V(\theta) \geq 2V(0)$ for all θ if $\mu\in(9/8,4]$. This follows
from the fact that $V(0)= -\mu^{-1/2} \leq -1/2 = V(\pi/2)/2 \leq V(\theta)/2$ for all θ if $\mu\in(9/8,4]$.

Therefore, the arc $\sigma_+^u(s)$ can only meet the curve w after crossing the circle $\{u=0\}$. Now, we shall study the crossing of $\sigma_+^u(s)$ with $\{u=0\}$.

Since the curve $\sigma_+^u(s)$ spirals around $(\pi/2,0)$ when $s \to +\infty$, it must cut u=0.
Let s* be the smallest value of s such that $u_{s*}(0)=0$. By Lemma 2, we have that
$u_{s*}(\tau)\geq 0$ for all $\tau< 0$. We shall prove that $\pi/2< \theta_{s*}(0)<\pi$. Since u'>0 in the segment $\{(\theta,u): 0<\theta< \pi/2$ and $u=0\}$, it follows that $\pi/2<\theta_{s*}(0)$. If $\theta_{s*}(0)>\pi$ then,
by using symmetry S_3, we have $\sigma_+^u(p^+(0),\mu)\cap\sigma_+^u(p^+(\pi),\mu)\neq\emptyset$, and this is not

possible. Of course, $\sigma_+^u(s*)$ is different from $(\pi/2,0)$ and $(\pi,0)$ because these two points belong to the homothetic orbits $\gamma_h(\pi/2)$ and $\gamma_h(\pi)$, respectively. So, $\pi/2 < \theta_{s*}(0) < \pi$.

Now, by symmetries (see Lemma 1) we have that the curves $\sigma_+^u(p^+(0),\mu)$, $\sigma_-^s(p^-(0),\mu)$, $\sigma_+^s(p^-(\pi),\mu)$ and $\sigma_-^u(p^+(\pi),\mu)$ look like those in Figure 6 for $0 < s < s*$. Again, by symmetries we have that $\sigma_+^u(p^+(0),\mu)$ is contained in the region bounded by the curve ABCDA, see Figure 6. Otherwise,
$\sigma_+^u(p^+(0),\mu) \cap \{\sigma_+^u(p^+(0),\mu) \cup \sigma_-^u(p^+(\pi),\mu)\} \neq \emptyset$ and this is not possible.

In short, Figure 5 follows.

Q.E.D.

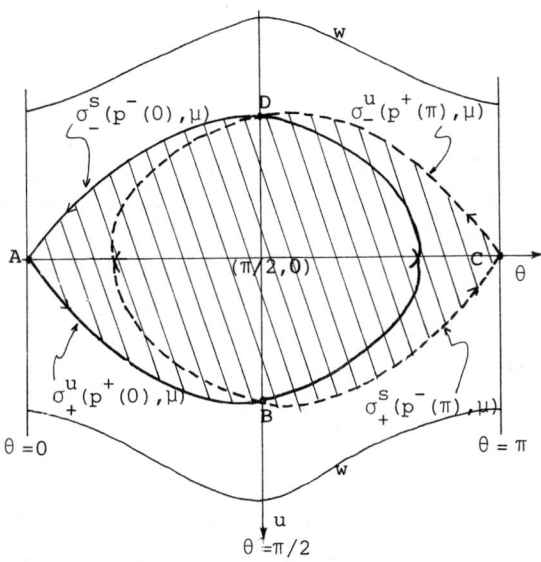

<u>Figure 6.</u> The curves $\sigma_-^s(p^-(0),\mu)$, $\sigma_+^u(p^+(0),\mu)$, $\sigma_+^s(p^-(\pi),\mu)$, $\sigma_-^u(p^+(\pi),\mu)$ and the points A,B,C and D.

In order to study the case $\mu>4$, let s' be the smallest value of $s\in[0,+\infty)$ such that $\sigma_+^u(s) \cap w = \emptyset$ for all $s\in(s',+\infty)$. By Proposition 5.3 of [D2] s' exists for all $\mu>9/8$. Of course, s' depends on μ. Let s" be the smallest value of $s\geq s'$ such that $\sigma_+^u(s) \cap \{(\theta,u) : 0\leq\theta< \pi/2 \text{ and } u=0\} \neq \emptyset$. Theorem 3 says that s'=s"=0 if $\mu\in[9/8,4]$; that is, $\sigma_+^u(s) \cap w = \emptyset$ for $\mu\in[9/8,4]$. Numerical computations show that $\sigma_+^u(s) \cap w = \emptyset$ until a value of μ, $\mu_c\in(4.9,5.0)$. See Appendix and Figures 7 for more numerical information on the continuity of the arc $\sigma_+^u(s)$.

PROPOSITION 4. _If_ $\mu > \mu_c$ _then_ $\sigma_+^u(s)$ _is a continuous arc for all_ $s \in [s'', \infty)$ _contained in_ $\{v' < 0\} \cap S$. _Furthermore,_ $\theta_s(0) \in [0, \pi)$ _for all_ $s \in [s'', \infty)$.

Proof. By definition, the continuity of the arc $\sigma_+^u(s)$ follows when $s \in [s'', \infty)$. Similar arguments used in the proof of Theorem 3 show that $\theta_s(0) \in [0, \pi)$ for all $s \in [s'', \infty)$.

                                                                        Q.E.D.

From now on, the arc $\sigma_+^u(s)$ or $\sigma_+^u(p^+(0), \mu)$ shall mean $\{\sigma_+^u(s) : s \geq s''\}$ if $\mu \geq \mu_c$ and $\{\sigma_+^u(s) : s \geq 0\}$ if $\mu \in [9/8, \mu_c)$. Similar notation is used for the other seven arcs $\sigma_{+,-}^{u,s}(p, \mu)$. These arcs can be defined by using the symmetries of Lemma 1.

LEMMA 5. _For_ $\mu > 1$ _we have that every orbit determined by a point of_ $\{v' < 0\} \cap S$ _either it tends to an equilibrium point on_ Λ, _or it meets in forward time the annulus_ S _in the region_ $\{v' \geq 0\} \cap S$.

Proof. Let $p \in \{v' < 0\} \cap S$ and let $p(\tau) = (r(\tau), v(\tau), \theta(\tau), u(\tau))$ be the solution determined by p for $\tau = 0$, i.e $p(0) = p$. Suppose that $v(\tau) < 0$ for all $\tau > 0$; that is, $r'(\tau) < 0$ for all $\tau > 0$. Then there exists $r(\infty) = \lim_{\tau \to \infty} r(\tau) \geq 0$. If $r(\infty) > 0$ then $v(\tau)$ will tend to zero when $\tau \to +\infty$, but this is impossible because $p(\tau)$ would tend to an equilibrium point out of the collision manifold. If $r(\infty) = 0$ then $p(\tau)$ tends to an equilibrium point.

                                                                        Q.E.D.

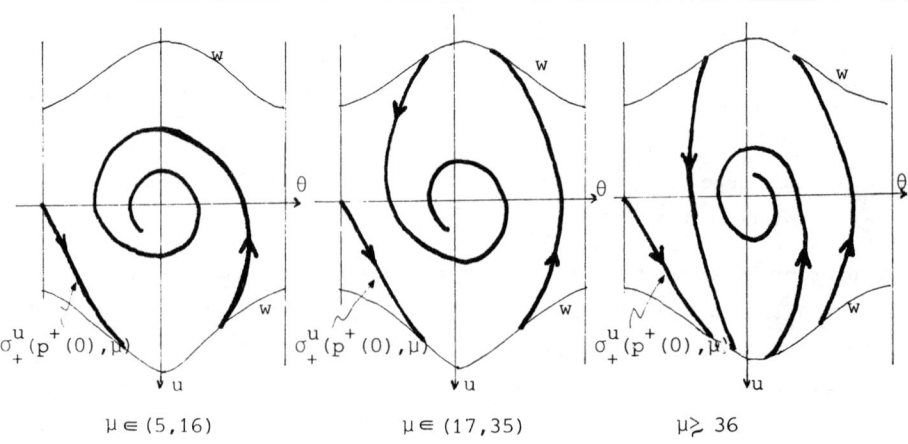

$\mu \in (5, 16)$ $\mu \in (17, 35)$ $\mu \geq 36$

Figures 7. The arc $\sigma_+^u(p^+(0), \mu)$ for a diferent values of μ.

(IV.2) The Poincaré maps g,f and h on v=0.

By Lemma 5, the orbit determined by a point
$p \in \{ v'<0 \} \cap S \setminus \{ \sigma_-^s(p^-(0),\mu) \cup \sigma_+^s(p^-(\pi),\mu) \cup \sigma_+^s(p^-(0),\mu) \cup \sigma_-^s(p^-(\pi),\mu) \}$ in forward
time always meets the annuli $S \cap \{v' \geq 0\}$ in a point p'. The map f(p)=p' is cal-
led the Poincaré map in forward time on $\{v'<0\} \cap S$.

Similarly, we can define a Poincaré map in backward time g on $\{v'<0\} \cap S$.
In fact, the domain of definition of g is,
$\{v'<0\} \cap S \setminus \{ \sigma_+^u(p^+(0),\mu) \cup \sigma_-^u(p^+(\pi,\mu) \cup \sigma_-^u(p^+(0),\mu) \cup \sigma_+^u(p^+(\pi),\mu) \}$.

We define two neighbourhoods $U_{+,-}^{s,u}(p,\mu)$ and $V_{+,-}^{s,u}(p,\mu)$ for each $\sigma_{+,-}^{s,u}(p,\mu)$ as
in Figure 8. Furthermore, for θ fixed there are always points of U farther
away from θ-axis than the points of V. The choice of these sixteenth neigbour-
hoods is made preserving the symmetries given in Lemma 1. For example, in
Figure 8 we have that, $U_-^u(p^+(\pi),\mu) = S_2 \circ S_0(U_+^u(p^+(0),\mu))$ and
$V_-^u(p^+(\pi),\mu) = S_2 \circ S_0(V_+^u(p^+(0),\mu))$.

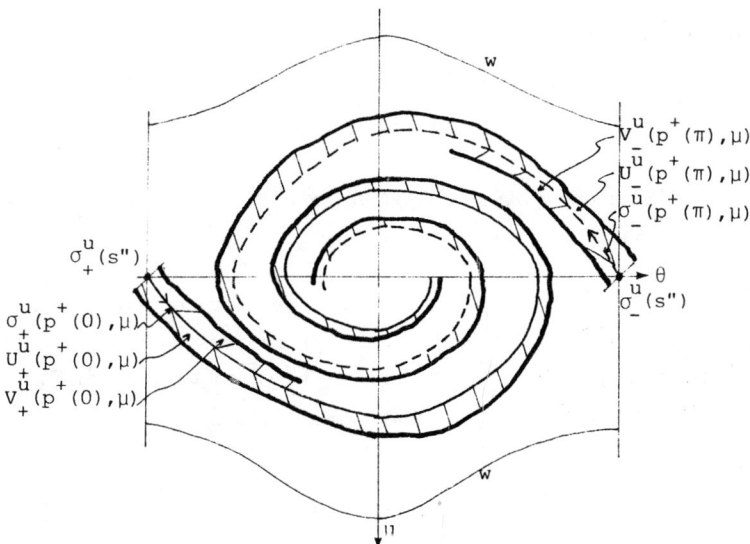

Figure 8. The neighbourhoods $U_{+,-}^{s,u}(p,\mu)$ and $V_{+,-}^{s,u}(p,\mu)$. Here, we have that
$\sigma_+^u(s'')=(0,0)$ for $\mu \in (9/8,4]$ and $\sigma_+^u(s'')$ is defined just before
to Proposition 4 for $\mu > 4$.

Also, we define the arcs γ, σ, σ', φ, φ' and the open regions X_i for i=1,2,3,4 as in Figure 9.

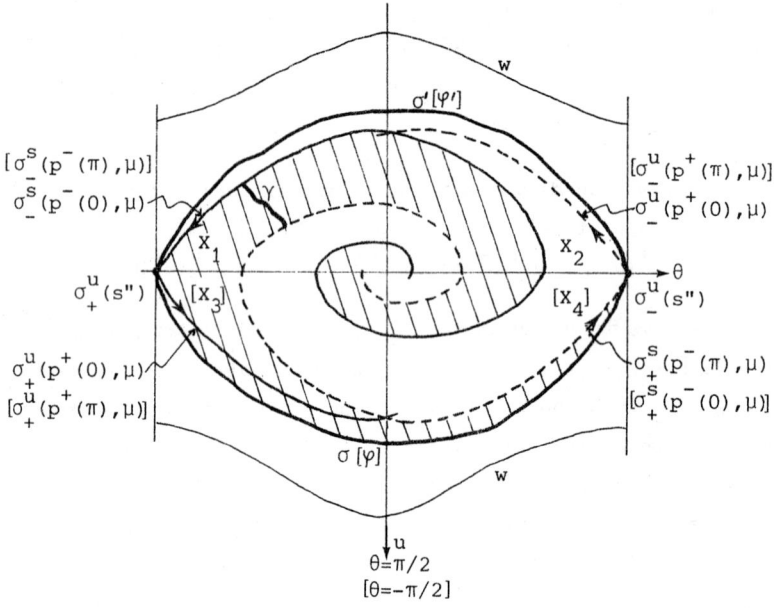

Figure 9. The arcs γ, σ, σ', φ, φ' and the regions X_i for i=1,2,3,4.

LEMMA 6. *$f(\gamma)$ is an open arc with endpoints $P_+^u(p^-(0),\mu)$ and $P_+^u(p^-(\pi),\mu)$ contained in $\{v' \geq 0\} \cap S$.*

Proof. By Lemma 5, we have that $f(\gamma)$ is an open arc contained in $\{v' \geq 0\} \cap S$. By definition of γ, Figures II.13 and II.14a, it follows that the endpoints of $f(\gamma)$ are either $P_+^u(p^-(0),\mu)$ and $P_+^u(p^-(\pi),\mu)$, or $P_-^u(p^-(0),\mu)$ and $P_-^u(p^-(\pi),\mu)$. Therefore, by using Figure II.14a the lemma follows.

Q.E.D.

By Lemma 6, $f(X_1)$ is the connected region bounded by $f(\sigma)$ and $\Lambda \cap S$, see Figures 10. Figure 10b is the same as Figure 10a. In 10b we take θ as an angular coordinate and u as a radial coordinate with u=0 in the circle shown in the picture. From now on, we shall use this type of representation for the annulus S.

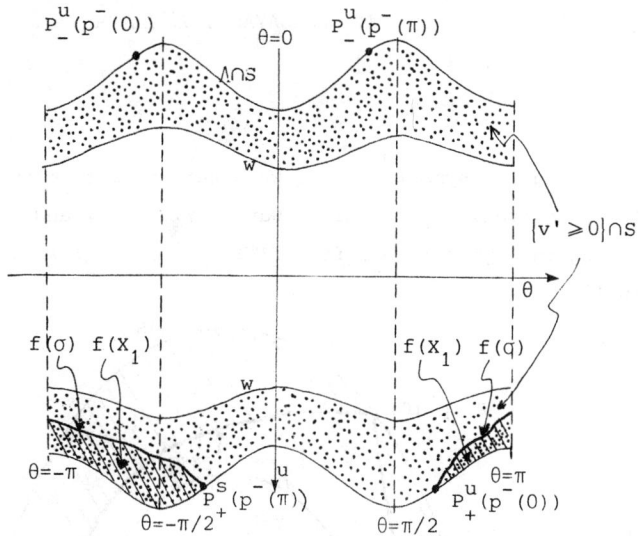

Figure 10a. The region $f(X_1)$.

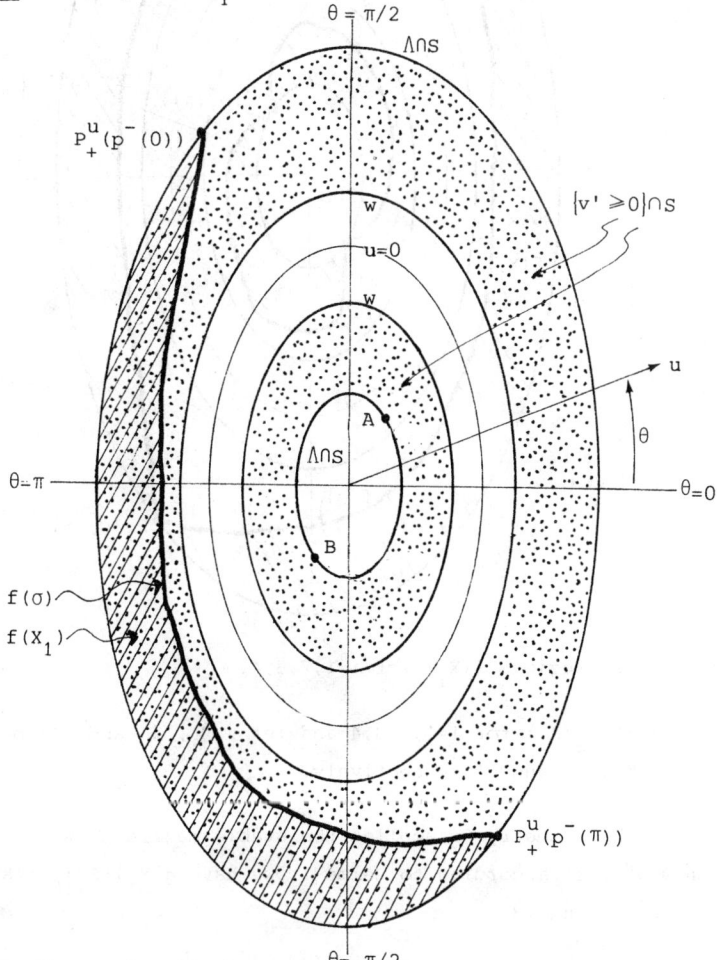

Figure 10b. The region $f(X_1)$.
Here $A = P_-^u(p^-(\pi))$ and $B = P_-^u(p^-(0))$.

By using the symmetry $S_2 \circ S_0$ we obtain from $f(\sigma)$ and $f(X_1)$, $f(\sigma')$ and $f(X_2)$ respectively. In a similar way, $f(\varphi)$, $f(\varphi')$ and $f(X_3)$, $f(X_4)$ are obtained, using the symmetry S_3, from $f(\sigma)$, $f(\sigma')$ and $f(X_1)$, $f(X_2)$ repectively, see Figure 11.

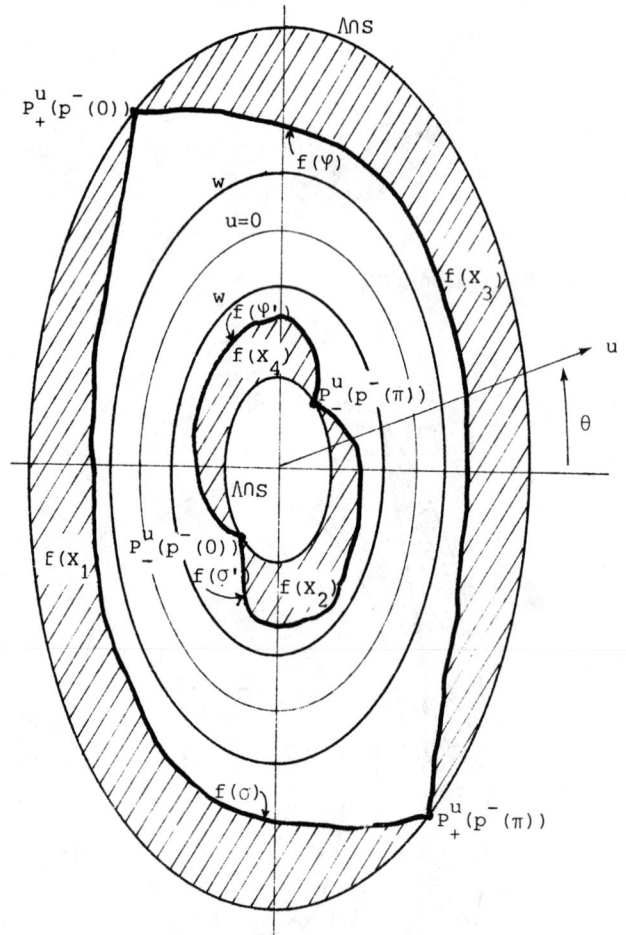

Figure 11. The regions $f(X_i)$ for $i=1,2,3,4$.

Let $Y_i = S_2|_S(X_i)$ for $i=1,2,3,4$ and let ψ, η, ψ' and η' be the images under $S_2|_S$ of σ, φ, σ' and φ' respectively.

Since the symmetry S_2 applies the stable manifolds $W^s(.,\mu)$ in the unstables ones $W^u(.,\mu)$ according to Lemma 1, we have $g(Y_i) = S_2|_S(f(X_i))$ for $i=1,2,3,4$ see Figures 12 and 13.

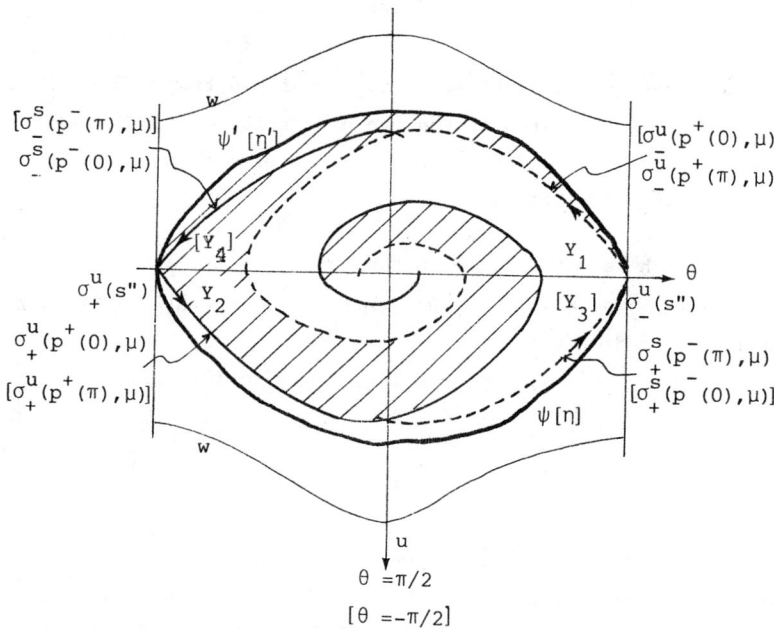

Figure 12. The arcs ψ, η, ψ', η' and the regions Y_i for $i=1,2,3,4$.

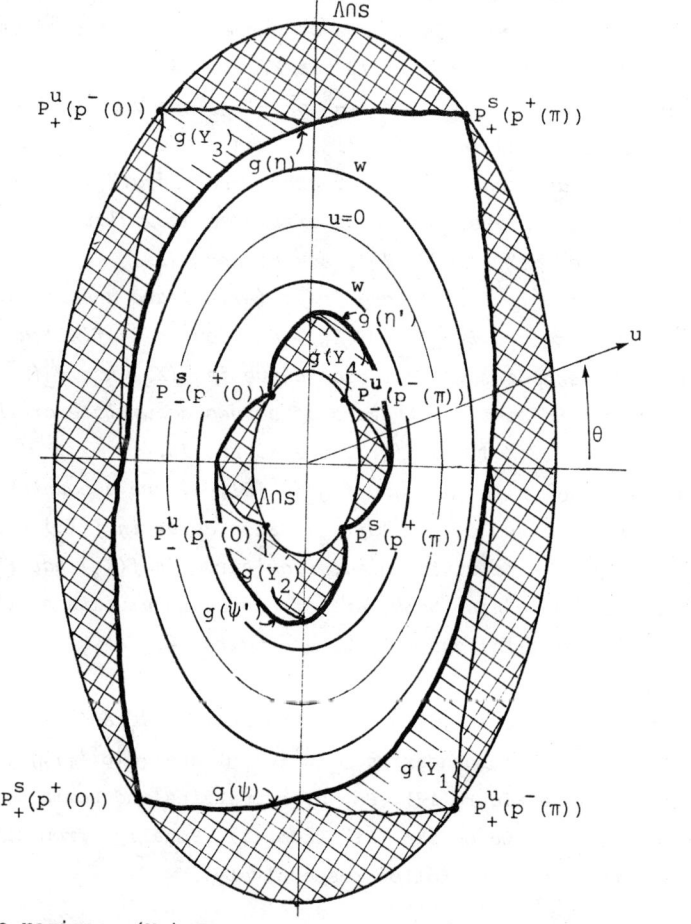

Figure 13. The regions $g(Y_i)$ for $i=1,2,3,4$ and $g(Y_j) \cap f(X_i)$ for $i,j=1,2,3,4$. The last ones are the shadowed regions ($\cancel{\boxtimes}$).

We consider the Poincaré map $h = f \circ g^{-1}$ defined from $\{v' \geq 0\} \cap S$ to itself. Actually, we are interested in the map h restricted to $g(Y_j) \cap f(X_i)$ for $i, j = 1, 2, 3, 4$; see Figure 13. Note that this intersection is not empty only if i and j have the same parity.

Note that h is not defined on the set,
$$g(\sigma_+^s(p^-(\pi), \mu) \cup \sigma_-^s(p^-(0), \mu) \cup \sigma_+^s(p^-(0), \mu) \cup \sigma_-^s(p^-(\pi), \mu)).$$ Similarly, the map h^{-1} is not defined on the set
$$f(\sigma_+^u(p^+(0), \mu) \cup \sigma_-^u(p^+(\pi), \mu) \cup \sigma_+^u(p^+(\pi), \mu) \cup \sigma_-^u(p^+(0), \mu)).$$

(IV.3) The invariant manifolds under g and f.

LEMMA 7. For $\mu > 9/8$ the following four statements hold:

(i) The sets $g(\sigma_+^s(p^-(\pi), \mu))$ and $g(\sigma_-^s(p^-(0), \mu))$ are a countable union of disjoint curves contained in $g(Y_1) \cup g(Y_2)$. Each curve in $g(Y_1)$ has $P_+^s(p^+(0), \mu)$ and $P_+^s(p^+(\pi), \mu)$ as endpoints. Also, each curve in $g(Y_2)$ has $P_-^s(p^+(0), \mu)$ and $P_-^s(p^+(\pi), \mu)$ as endpoints. Both sets of curves accumulate at the collision boundaries $\Lambda \cap S$ of $g(Y_1)$ and $g(Y_2)$, see Figure 14.

(ii) The sets $g(\sigma_+^s(p^-(0), \mu)$ and $g(\sigma_-^s(p^-(\pi), \mu)$ are a countable union of disjoint curves contained in $g(Y_3) \cup g(Y_4)$. Each curve in $g(Y_3)$ has $P_+^s(p^+(\pi), \mu)$ and $P_+^s(p^+(0), \mu)$ as endpoints. Also, each curve on $g(Y_4)$ has $P_-^s(p^+(0), \mu)$ and $P_-^s(p^+(\pi), \mu)$ as endpoints. Both sets of curves accumulate at the collision boundaries $\Lambda \cap S$ of $g(Y_3)$ and $g(Y_4)$, see Figure 14.

(iii) The sets $f(\sigma_+^u(p^+(0), \mu))$ and $f(\sigma_-^u(p^+(\pi), \mu))$ are a countable union of disjoint curves contained in $f(X_1) \cup f(X_2)$. Each curve in $f(X_1)$ has $P_+^\mu(p^-(0), \mu)$ and $P_+^\mu(p^-(\pi), \mu)$ as endpoints. Also, each curve in $f(X_2)$ has $P_-^\mu(p^-(0), \mu)$ and $P_-^\mu(p^-(\pi), \mu)$ as endpoints. Both sets of curves accumulate at the collision boundaries $\Lambda \cap S$ of $f(X_1)$ and $f(X_2)$, see Figure 15.

(iv) The sets $f(\sigma_+^u(p^+(\pi), \mu))$ and $f(\sigma_-^u(p^+(0), \mu))$ are a countable union of disjoint curves contained in $f(X_3) \cup f(X_4)$. Each curve in $f(X_3)$ has $P_+^\mu(p^-(\pi), \mu)$ and $P_+^\mu(p^-(0), \mu)$ as endpoints. Also, each curve in $f(X_4)$ has $P_-^\mu(p^-(\pi), \mu)$ and $P_-^\mu(p^-(0), \mu)$ as endpoints. Both sets of curves accumulate at the collision boundaries $\Lambda \cap S$ of $f(X_3)$ and $f(X_4)$, see Figure 15.

Proof. Note that $\sigma_+^s(p^-(\pi), \mu)$ crosses the regions Y_1 and Y_2 alternatively, each time cutting the boundaries $\sigma_+^u(p^+(0), \mu)$ and $\sigma_-^u(p^+(\pi), \mu)$, and tends to the point $(\pi/2, 0)$ (see Figure 12). So, by symmetries and Lemma 6, (i) follows.

By symmetry S_3 we obtain (ii) from (i). Finally, from (i), (ii) and symmetry S_2, (iii) and (iv) follow respectively.

Q.E.D.

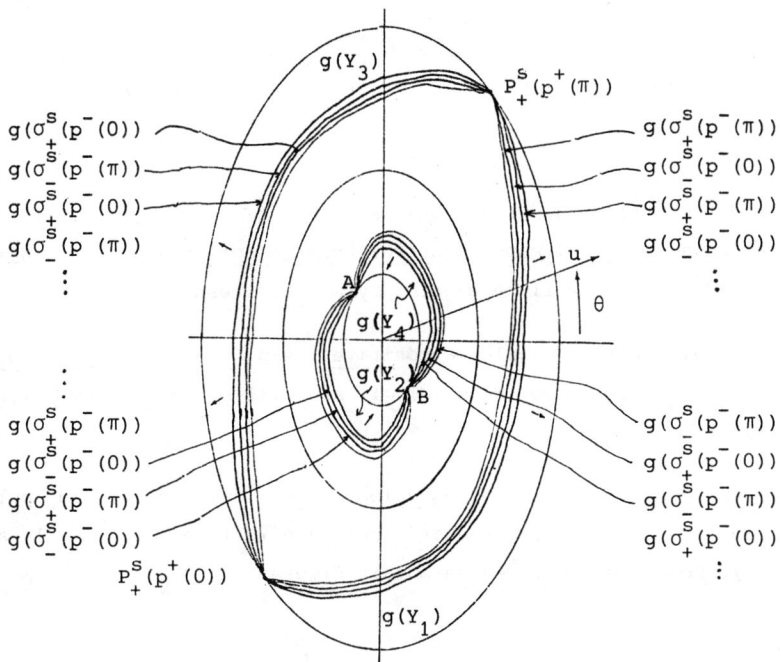

Figure 14. The sets $g(\sigma^{S}_{+,-}(p^{-}(\theta_{o}),\mu)$ for $\theta_{o} = 0$ or π. Here $A = P^{S}_{-}(p^{+}(0))$ and $B = P^{S}_{-}(p^{+}(\pi))$

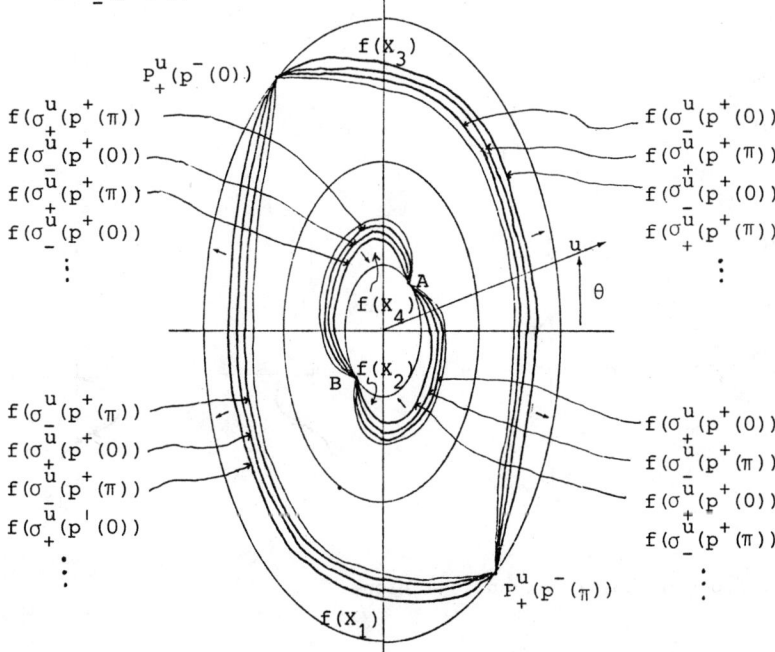

Figure 15. The sets $f(\sigma^{u}_{+,-}(p^{+}(\theta_{o}),\mu)$ for $\theta_{o} = 0$ or π. Here $A = P^{u}_{-}(p^{-}(\pi))$ and $B = P^{u}_{-}(p^{-}(0))$.

(IV.4) Geometrical interpretation of the neighbourhoods of the invariant
      manifolds.

We need the following definitions. An orbit $p(\tau) = (r(\tau), \theta(\tau), v(\tau), u(\tau))$
will be called a <u>positive</u>(resp. <u>negative</u>) <u>upper</u> <u>ejection</u> when $p(\tau) \in W_-^u(p^+(\pi), \mu)$
(resp. $W_+^u(p^+(0), \mu)$) and will be denoted by $e(+,u)$ (resp. $e(-,u)$).

An orbit $p(\tau)$ will be called a <u>positive</u> (resp. <u>negative</u>) <u>upper</u> <u>collision</u>
when $p(\tau) \in W_-^s(p^-(0), \mu)$ (resp. $W_+^s(p^-(\pi), \mu)$) and will be denoted by $c(+,u)$
(resp. $c(-,u)$).

A <u>positive</u> or <u>negative</u> <u>lower</u> <u>ejection</u> or <u>collision</u> is defined applying
S_3 to the above definitions, we denote them by $e(+,l)$, $e(-,l)$, $c(+,l)$ and $c(-,l)$
respectively. From Figure II.14a we obtain Figure 16.

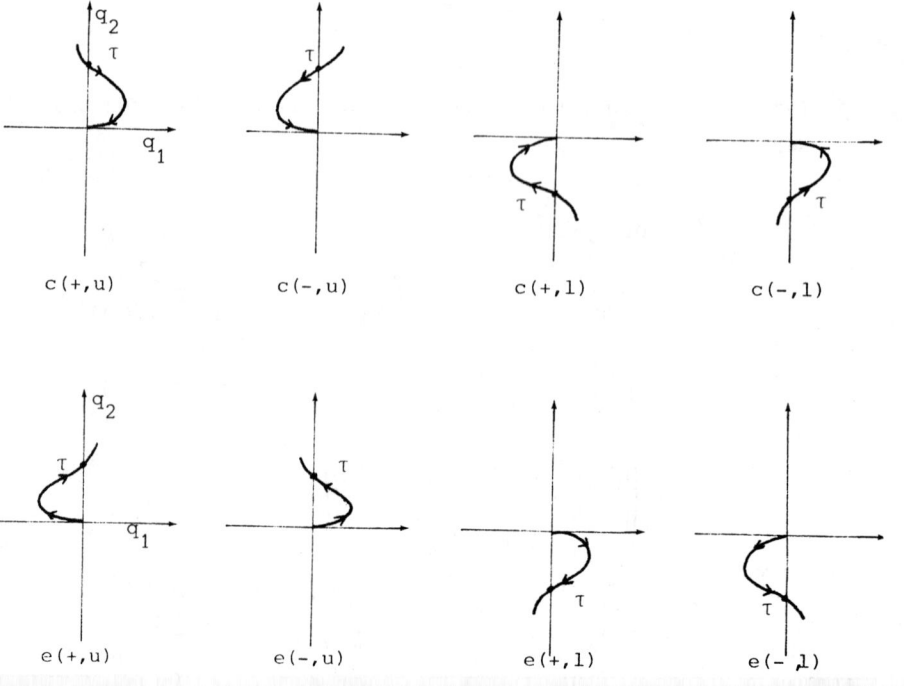

Figure 16. Types of collisions and ejections.

We note that Lemma 7 gives us the geometrical location of these ejection
and collision orbits on the boundary of the domain of definition of h and h^{-1}.

An orbit $p(\tau)$ will be called a <u>positive</u> (resp. <u>negative</u>) <u>upper</u> <u>crossing</u> when there are $\tau_1 < \tau_2 < \tau_3 < \tau_4$ such that $\theta(\tau_1) = \theta(\tau_2) = \pi/2$, $\theta(\tau_3) = \theta(\tau_4) = -\pi/2$, $\theta(\tau) \neq \pm\pi/2$ for all $\tau \in (\tau_1,\tau_2) \cup (\tau_2,\tau_3) \cup (\tau_3,\tau_4)$, and $\theta(\tau)$ for $\tau \in (\tau_2,\tau_3)$ turns clockwise (resp. counterclockwise). We will denote it by $C(+,u)$ (resp. $C(-,u)$).

An orbit $p(\tau)$ will be called a <u>positive</u> (resp.<u>negative</u>) <u>upper</u> <u>rotation</u> when there are $\tau_1 < \tau_2 < \tau_3$ such that $\theta(\tau_1) = \theta(\tau_3) = \pi/2$, $\theta(\tau_2) = -\pi/2$, $\theta(\tau) \neq \pm\pi/2$ for all $\tau \in (\tau_1,\tau_2) \cup (\tau_2,\tau_3)$, and $\theta(\tau)$ for $\tau \in (\tau_1,\tau_3)$ turns clockwise (resp. counterclockwise). We will denote it by $R(+,u)$ (resp. $R(-,u)$).

A <u>positive</u> or <u>negative</u> <u>lower</u> <u>crossing</u> or <u>rotation</u> is defined applying S_3 to the above definitions, we denote them by $C(+,l)$, $C(-,l)$, $R(+,l)$ and $R(-,l)$ respectively, see Figure 17.

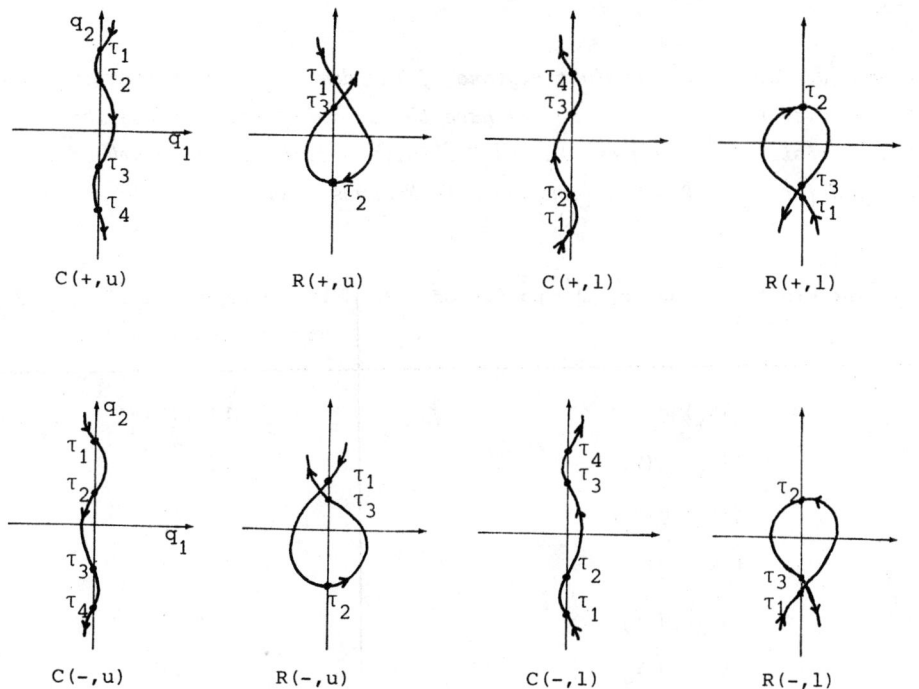

C(+,u)   R(+,u)   C(+,l)   R(+,l)

C(-,u)   R(-,u)   C(-,l)   R(-,l)

<u>Figure 17</u>. Types of crossings and rotations.

<u>THEOREM 8</u>. *For $\mu > 9/8$ the eight regions of Table 1 contained in the domain of the definition of h (see the shadowed regions of Figure 13) have the dynamical behaviour described in Table 1.*

The orbits defined by the points of	Have a dynamical behaviour
$f(U^s_+(p^-(\pi),\mu) \cap g(U^u_+(p^+(0),\mu)) \subset f(X_1) \cap g(Y_1)$	R(-,u)
$f(V^s_-(p^-(0),\mu)) \cap g(V^u_-(p^+(0),\mu)) \subset f(X_1) \cap g(Y_3)$	C(-,u)
$f(V^s_-(p^-(\pi),\mu)) \cap g(V^u_-(p^+(\pi),\mu)) \subset f(X_3) \cap g(Y_1)$	C(-,1)
$f(U^s_+(p^-(0),\mu)) \cap g(U^u_+(p^+(\pi),\mu)) \subset f(X_3) \cap g(Y_3)$	R(-,1)
$f(U^s_-(p^-(0),\mu)) \cap g(U^u_-(p^+(\pi),\mu)) \subset f(X_2) \cap g(Y_2)$	R(+,u)
$f(V^s_+(p^-(\pi),\mu)) \cap g(V^u_+(p^+(\pi),\mu)) \subset f(X_2) \cap g(Y_4)$	C(+,u)
$f(V^s_+(p^-(0),\mu)) \cap g(V^u_+(p^+(0),\mu)) \subset f(X_4) \cap g(Y_2)$	C(+,1)
$f(U^s_-(p^-(\pi),\mu)) \cap g(U^u_-(p^+(0),\mu)) \subset f(X_4) \cap g(Y_4)$	R(+,1)

Table 1.

THEOREM 9. *For $\mu > 9/8$ the eight regions of the domain of definition of h (see the shadowed regions of Figure 13) have the dynamic behaviour described in Table 2, when we restrict the regions X_i and Y_i, $i=1,2,3,4$, to a neighbourhood U of $\gamma_h(\theta_o) \cap S$ for $\theta_o = \pi/2, -\pi/2$ (compare with Figures 9 and 12)*

The orbits defined by the points of	In a neighbourhood of $\gamma_h(\theta_o)$ S have a dynamical behaviour
$f(X_1) \cap g(Y_1)$	R(-,u)
$f(X_1) \cap g(Y_3)$	C(-,u)
$f(X_3) \cap g(Y_1)$	C(-,1)
$f(X_3) \cap g(Y_3)$	R(-,1)
$f(X_2) \cap g(Y_2)$	R(+,u)
$f(X_2) \cap g(Y_4)$	C(+,u)
$f(X_4) \cap g(Y_2)$	C(+,1)
$f(X_4) \cap g(Y_4)$	R(+,1)

Table 2.

The proofs of Theorem 8 and 9 are similar in the sense that we use the behaviour of the invariant manifolds. Therefore, we only prove Theorem 9.

Proof (Theorem 9). Figure 13 describes qualitatively the eight regions
$f(X_i) \cap g(Y_j)$ when X_i and Y_j (i,j=1,2,3,4) are defined in a convenient and
sufficiently small neighbourhood U of $\gamma_h(\theta_0) \cap S$ for $\theta_0 = \pi/2$, $-\pi/2$. We consider
an orbit $p(\tau) = (r(\tau), v(\tau), \theta(\tau), u(\tau))$ such that $p(0) \in f(X_1) \cap g(Y_1)$. From Figure 13 and Lema II.11 it follows that $\pi < \theta(0) < 2\pi$ and $u(0) > 0$.

Let τ_1 be the smallest value of $\tau > 0$ such that $u(\tau)=0$. We claim that τ_1
exists and $\pi/2 < \theta(\tau_1) < \pi$. In order to prove the claim we take $g^{-1}(p(0)) \in Y_1 \cap U$;
of course, $g^{-1}(p(0)) = p(\tau_2)$ with $\tau_2 > 0$. Since the flow of $Y_1 \cap U$ in backward time
firstly follows near $\gamma_h(\pi/2)$ and after near Λ, we have that there exists
$\tau \in (0, \tau_2)$ such that $u(\tau)=0$. Then there exists $\tau_1 \in (0, \tau_2)$ such that τ_1 is the
smallest $\tau > 0$ with $u(\tau_1)=0$. From the behaviour of the flow in forward time of
$B_+^s(p^+(0), \mu)$ and $B_+^u(P^-(\pi), \mu)$ (see Figure II.13 and II.14a), we have that
$\pi/2 < \theta(\tau_1) < \pi$. So, the claim is proved.

Let τ_3 be the largest value of $\tau < 0$ such that $u(\tau)=0$. We shall prove that
τ_3 exists and $0 < \theta(\tau_3) < \pi/2$. We denote by $p(\tau_4)$ the point $f^{-1}(p(0)) \in X_1 \cap U$; of
course $\tau_4 < 0$. Since the flow of $X_1 \cap U$ in forward time firstly follows close to
$\gamma_h(\pi/2)$ and after near Λ, we have that there exists $\tau_3 \in (\tau_4, 0)$ such that τ_3 is
the largest $\tau < 0$ with $u(\tau_3)=0$. From the behaviour of the flow in backward time
of $B_+^s(p^+(0), \mu)$ and $B_+^u(p^-(\pi), \mu)$ (see Figures II.13 and II.14a), we have that
$0 < \theta(\tau_3) < \pi/2$.

Now, we suppose that $\pi < \theta(0) < 3\pi/2$ (the case $3\pi/2 \leq \theta(0) < 2\pi$ follows simi-
larly). Since $\pi/2 < \theta(\tau_1) < \pi$, $0 < \theta(\tau_3) < \pi/2$ and $u(\tau) = \theta'(\tau) > 0$ for all $\tau \in (\tau_3, \tau_1)$,
the orbit $p(\tau)$ for $\tau \in (\tau_3, \tau_1)$ looks like the one in Figure 18. So by continuity
of $\theta(\tau)$ there exists $\tau_1' \in (\tau_3, 0)$, $\tau_2' \in (0, \tau_1)$ and $\tau_3' \in (\tau_2', \tau_1)$ such that
$\theta(\tau_1') = \theta(\tau_3') = \pi/2$ and $\theta(\tau_2') = -\pi/2$. Hence, the orbit $p(\tau)$ has a negative upper
rotation, $R(-, u)$.

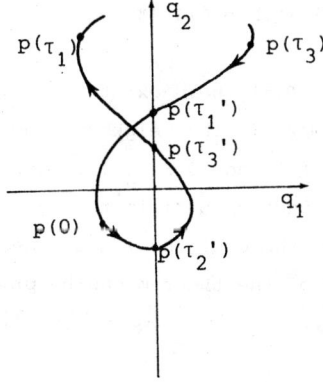

Figure 18. The orbit $p(\tau)$ with $p(0) \in f(X_1) \cap g(Y_1)$.

Now, we consider an orbit $p(\tau)$ such that $p(0) \in f(X_1) \cap g(Y_3)$. From Figure 13 it follows that $\pi/2 < \theta(0) < 3\pi/2$ and $u(0) > 0$. As in the above case there exists τ_3 the largest value of $\tau < 0$ such that $u(\tau) = 0$ and $0 < \theta(\tau_3) < \pi/2$.

Let τ_1 be the smallest value of $\tau > 0$ such that $u(\tau) = 0$. We shall prove that τ_1 exists and $-\pi/2 < \theta(\tau_1) < 0$. We denote by $p(\tau_2)$ the point $g^{-1}(p(0)) \in Y_3 \cap U$, so $\tau_2 > 0$. Since the flow of $Y_3 \cap U$ in backward time firstly follows near $\gamma_h(-\pi/2)$ and after near Λ, we have that there exists $\tau \in (0, \tau_2)$ such that $u(\tau) = 0$. Then there exists $\tau_1 \in (0, \tau_2)$ such that τ_1 is the smallest $\tau > 0$ with $u(\tau) = 0$. From the behaviour of the flow in forward time of $B_+^s(p^+(0), \mu)$ and $B_+^u(p^-(0), \mu)$ (see Figures II.13 and II.14a) we have that $-\pi/2 < \theta(\tau_1) < 0$.

Now, we suppose that $\pi/2 < \theta(0) < \pi$ (the case $\pi \leq \theta(0) < 3\pi/2$ follows similarly). Since $-\pi/2 < \theta(\tau_1) < 0$, $0 < \theta(\tau_3) < \pi/2$ and $u(\tau) > 0$ for all $\tau \in (\tau_3, \tau_1)$, the orbit of $p(\tau)$ for $\tau \in (\tau_3, \tau_1)$ looks like the one in Figure 19.

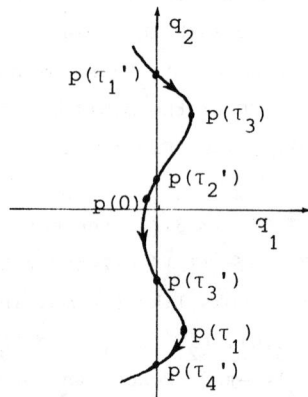

Figure 19. The orbit $p(\tau)$ with $p(0) \in f(X_1) \cap g(Y_3)$.

So, by continuity of $\theta(\tau)$ there exist $\tau_2' \in (\tau_3, 0)$ and $\tau_3' \in (0, \tau_1)$ such that $\theta(\tau_2') = \pi/2$ and $\theta(\tau_3') = -\pi/2$. If the neighbourhood U is sufficiently small, then there exist $\tau_4' \in (\tau_1, \tau_2)$ and $\tau_1' \in (\tau_4, \tau_3)$ such that $\theta(\tau_4') = -\pi/2$ and $\theta(\tau_1') = \pi/2$, because the function $u(\tau)$ in the intervals (τ_1, τ_2) and (τ_4, τ_3) has more than one zero. Hence the orbit $p(\tau)$ has a negative upper crossing, $c(-, u)$.

The other six cases of the theorem can be proved either by using symmetries or in a similar way.

$$Q.E.D.$$

We note that if the neighbourhoods $U_{+,-}^{u,s}(p, \mu)$, $V_{+,-}^{u,s}(p, \mu)$ and the neighbourhood of $\gamma_h(\theta_o) \cap S$ of Theorems 8 and 9 are adequate, then Theorem 9 follows from Theorem 8.

(IV.5) Regions with a constant number of crossings with the q_2-axis,
the map $S_{(\theta_o,\mu)}$.

Let $U(\theta_o,\mu)$ be the shadowed region of Figure 20 for $\theta_o = \pm\,\pi/2$.

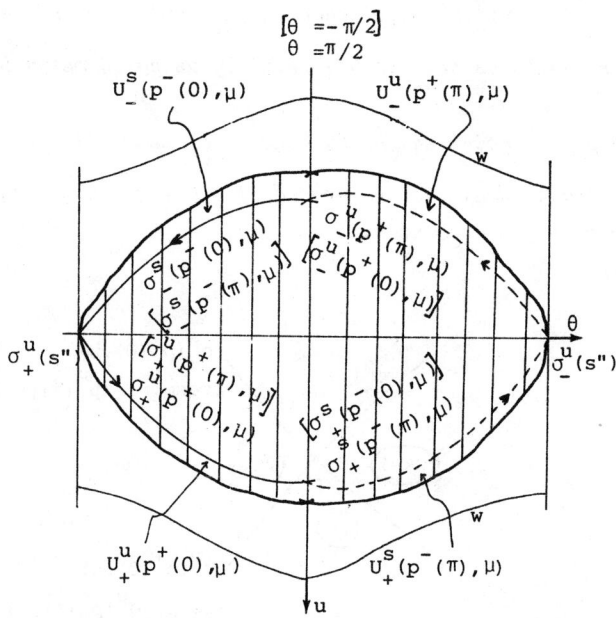

Figure 20. The regions $U(\theta_o,\mu)$ for $\theta_o = \pm\,\pi/2$.

For any $p \in U(\theta_o,\mu)$, $S_{(\theta_o,\mu)}(p)$ denotes the number of times the orbit through p crosses the q_2-axis between the crossings of the q_1-axis just prior to and just after p along the orbit. The ejection just prior to p and the collision just after p are also computed as a crossing .

The following three lemmas are due to Devaney, see [D5, p.303-304].

LEMMA 10. _(i) The map_ $S_{(\pi/2,\mu)}$ _defined on_ $U(\pi/2,\mu)$ _is continuous on_
$U(\pi/2,\mu) \backslash \left| \sigma^s_-(p^-(0),\mu) \cup \sigma^s_+(p^-(\pi),\mu) \cup \sigma^u_+(p^+(0),\mu) \cup \sigma^u_-(p^+(\pi),\mu) \right|$. _On disconti-_
nuity points $S_{(\pi/3,\mu)}$ _increases or decreases by 1. Also,_ $S_{(\pi/2,\mu)}(p) \to +\infty$
when $p \to (\pi/2,0)$.
(ii) A similar result is true for the map $S_{(-\pi/2,\mu)}$.

LEMMA 11. Let $p \in U(\pi/2, \mu)$ with $p \neq (\pi/2, 0)$. Then the following hold.

(i) If $p \in \sigma_-^s(p^-(o), \mu) \cap \sigma_+^u(p^+(0), \mu)$ then $S_{(\pi/2, \mu)}(p)$ is even.

(ii) If $p \in \sigma_-^s(p^-(0), \mu) \cap \sigma_-^u(p^+(\pi), \mu)$ then $S_{(\pi/2, \mu)}(p)$ is odd.

(iii) If $p \in \sigma_+^s(p^-(\pi), \mu) \cap \sigma_+^u(p^+(0), \mu)$ then $S_{(\pi/2, \mu)}(p)$ is odd.

(iv) If $p \in \sigma_+^s(p^-(\pi), \mu) \cap \sigma_-^u(p^+(\pi), \mu)$ then $S_{(\pi/2, \mu)}(p)$ is even.

A similar result is true in $U(-\pi/2, \mu)$ by using symmetry S_3.

LEMMA 12. Let p be a point in the hypotheses of Lemma 11. Then, the values of $S_{(\pi/2, \mu)}$ in a small enough neighbourhood of p are given in Figures 21. A similar result is true for a point $p \in U(-\pi/2, \mu)$.

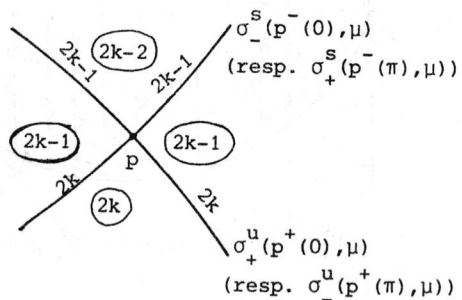

Figure 21a. This picture gives the values of $S_{(\pi/2, \mu)}$ in a neighbourhood of p in the case (i) (resp. (iv)) of Lemma 11 when $S_{(\pi/2, \mu)}(p) = 2k$.

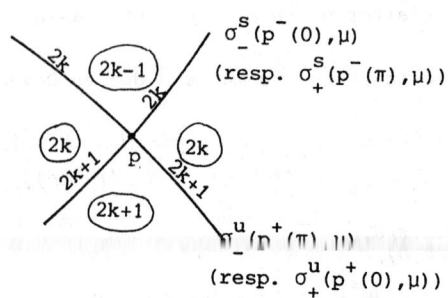

Figure 21b. This picture gives the values of $S_{(\pi/2, \mu)}$ in a neighbourhood of p in the case (ii) (resp. (iii)) of Lemma 11 when $S_{(\pi/2, \mu)}(p) = 2k+1$.

Proof(Lemma 10, 11 and 12). Lemma 11 follows from Figure 16, counting argument and the fact that the orbit through p can not be tangent to the q_2-axis.

From the local behaviour of the solutions near the ones which have a point on $\sigma^{s,u}_{+,-}(p^{\pm}(\theta_0),\mu)$ with $\theta_0=0,\pi$ (see Figure II.14a), we obtain lemmas 10 and 12.

Q.E.D.

THEOREM 13. *The following holds.*

$$S_{(\pm\pi/2,\mu)}(U(\pm\pi/2,\mu)) = \begin{cases} \{1,2,3,4,\ \ldots\ \} & if\ \mu\in[9/8,\mu_c) \\ \{2n_o-1,\ 2n_o,\ \ldots\ \} & if\ \mu\geqslant\mu_c\ , \end{cases}$$

where $n_o = n_o(\mu) \geq 1$.

Proof. By symmetries and Lemmas 10, 11 and 12, we obtain Figures 22. Figure 22b is a qualitative and realistic picture if we are in a small enough neighbourhood of $(\pm\pi/2,0)$. In fact, Devaney in [D2] has proved that if $\mu>9/8$ then, in a neighbourhood of $(\pm\pi/2,0)$, $\sigma^{u,s}_{+,-}(p(\theta_1,\mu))$ where $\theta_1=0,\pi$ are spirals. Figure 22a is a qualitative and perhaps realistic picture too, but at least it has the number of indicated regions (see Figure 23) when $\mu\in[9/8,\mu_c)$.

Note that the point p' in Figure 22a corresponds to an orbit as in Figure 24, then $S_{(\pi/2,\mu)}(p')=3$ and all of the enumeration of the different regions of Figure 22a follows using Lemma 12. If we have $S_{(\pi/2,\mu)}(p')= 2n_o+1$ for the point p' in Figure 22b then all the enumeration follows in the same way. This implies the theorem.

Q.E.D.

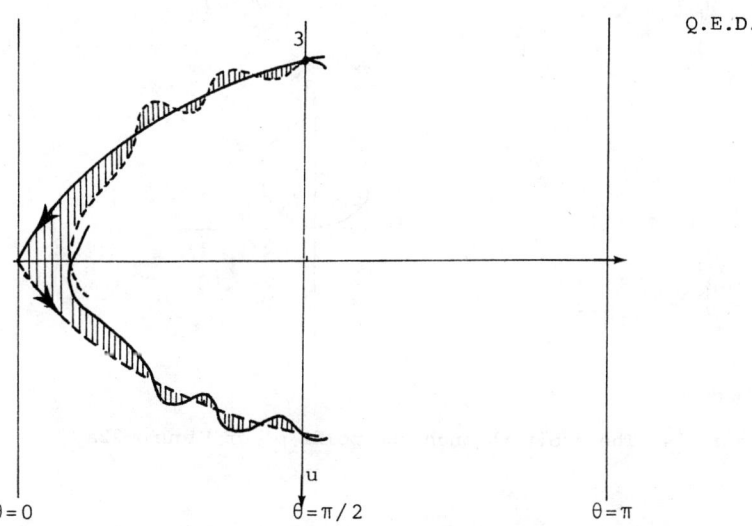

θ=0   θ=π/2   θ=π

Figure 23. In this case one of the region in Figure 22a has been broken in
           nine regions.

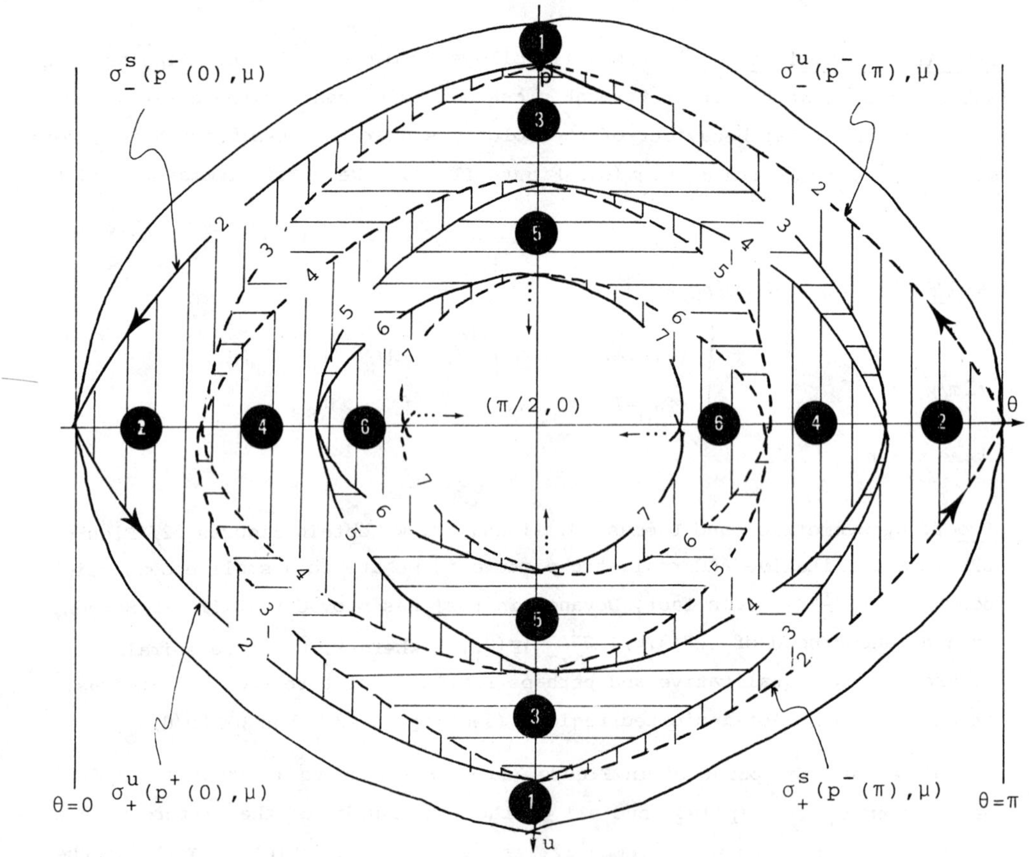

$\sigma_-^s(p^-(0),\mu)$
$\sigma_-^u(p^-(\pi),\mu)$
$\sigma_+^u(p^+(0),\mu)$
$\sigma_+^s(p^-(\pi),\mu)$
$(\pi/2,0)$
$\theta=0$
$\theta=\pi$
θ
u
p'

Figure 22a. The values of $S_{(\pi/2,\mu)}$ on $U(\pi/2,\mu)$. Here the numbers inside the circles are the values of $S_{(\pi/2,\mu)}$ for the open regions.

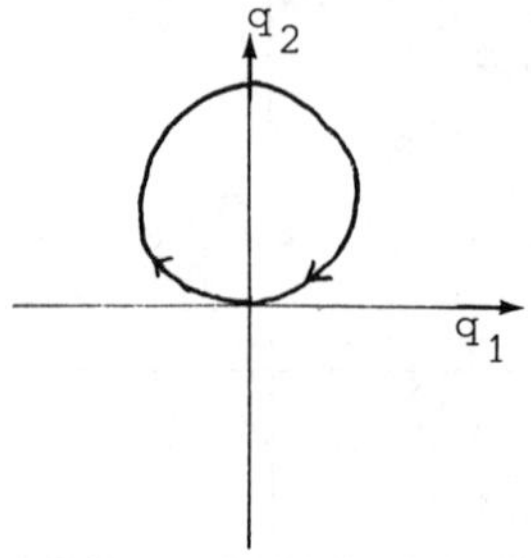

q_2
q_1

Figure 24. The orbit through the point p' of Figure 22a.

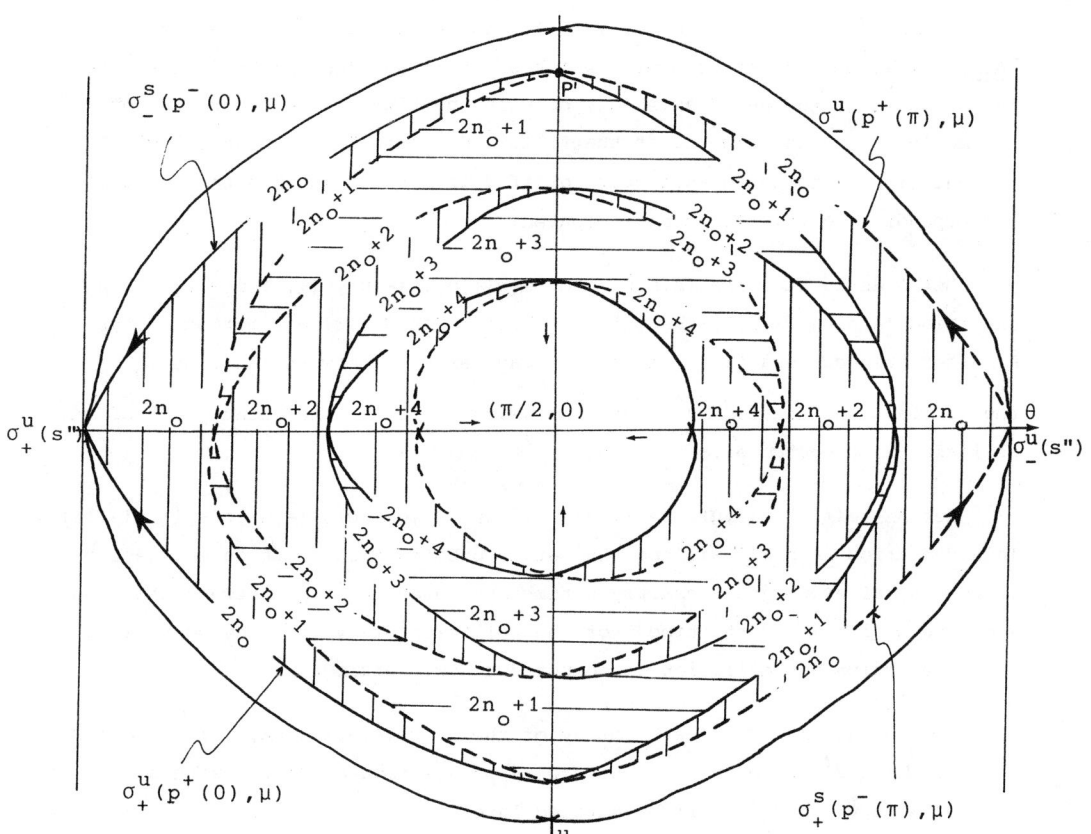

Figure 22b. The values of $S_{(\pi/2,\mu)}$ in a neighbourhood of $(\pi/2,0)$.

(IV.6) Basic sets for dynamical description.

Each one of the eight regions of the domain of definition of h given in Table 2, from now on, will be denoted by their dynamic behaviour. For instance, the region $f(X_1) \cap g(Y_1)$ in a neighbourhood U of $\gamma_h(\pi/2) \cap S$ will be denoted by $R(-,u)$.

In order to study the dynamic of the Poincaré map h we need some definitions.

Let $A,B \in \{C(+,u), C(-,u), R(+,u), R(-,u), C(+,1), C(-,1), R(+,1), R(-,1)\}$. Then the triad (A,n,B) will be the set of orbits which describe a motion of type A, after they cut exactly n-times the heavy axis q_2 between the time τ_3 of the motion A and the time τ_2 of the following motion B; if B is a rotation then the cut corresponding to time τ_2 is not taken into account.

Let $A \in \{e(+,u), e(-,u), e(+,l), e(-,l)\}$ and B as above. Then the triad [A,n,B) will be the set of orbits which start in an ejection of type A, after cut n times the heavy axis, q_2, between the time τ (see Figure 16) of the motion A and the time τ_2 of the following motion B. The point (0,0) of the ejection is computed as a cutting; if B is a rotation then the cutting corresponding to time τ_2 is not computed.

Let A as in the case (A,n,B) and B$\in\{c(+,u), c(-,u), c(+,l), c(-,l)\}$ .Then the triad (A,n,B] will be the set of orbits which describe a motion of type A, after they cut exactly n times the heavy axis q_2 between the time τ_3 of the motion A and the time τ of the following motion B. The point (0,0) of collision is computed as a cutting.

Let $A \in \{e(+,u), e(-,u), e(+,l), e(-,l)\}$ and B$\in\{c(+,u), c(-,u), c(+,l), c(-,l)\}$. Then the triad [A,n,B] will be the set of orbits which start in an ejection of type A, after they cut exactly n times the heavy axis q_2 between the time τ of the motion A and the time τ of the collision of type B. The point (0,0) of the ejection and collision is computed as two cuttings.

Now, we shall study the topology of the sets (A,n,B), [A,n,B) and(A,n,B]. The sets [A,n,B] correspond to the orbits of ejection-collision with a unique cut with v=0 and it will be studied later.

First of all we consider the region R(-,u), see Table 2. If the neighbourhood U of $\gamma_h(\theta_o) \cap S$ is sufficiently small, then from Figure 13 we have that the region R(-,u) is bounded by the curves a,b and $\Lambda \cap S$, see Figure 25. Therefore, the set $g^{-1}(R(-,u))$ is described in Figure 26.

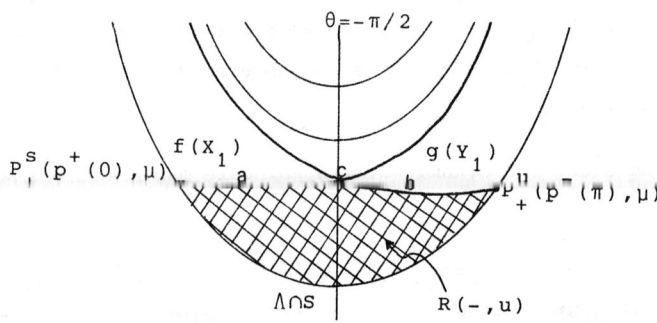

Figure 25. The region R(-,u).

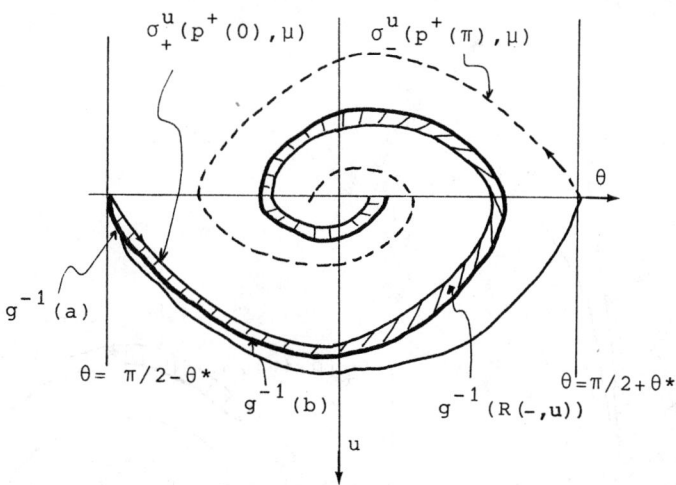

Figure 26. The set $g^{-1}(R(-,u))$. Here $0 < \theta^* = \theta^*(\mu, U)$ and U is the neighbourhood given by Theorem 9.

Again from Figure 13, we have that the region $C(-,u) = f(X_1) \cap g(Y_3)$ is bounded by the curves a, b and $\Lambda \cap S$ of Figure 27. Then the set $g^{-1}(C(-,u)) \subset Y_3$ is described in Figure 28.

We can study the other six cases of Table 2 either in a similar way or by using the symmetries as follows:

$$g^{-1}(R(-,1)) = S_3(g^{-1}(R(-,u))),$$
$$g^{-1}(C(-,1)) = S_3(g^{-1}(C(-,u))),$$
$$g^{-1}(R(+,1)) = S_2 \circ S_0(g^{-1}(R(-,1))),$$
$$g^{-1}(C(+,1)) = S_2 \circ S_0(g^{-1}(C(-,1))),$$
$$g^{-1}(R(+,u)) = S_3(g^{-1}(R(+,1)) \text{ and}$$
$$g^{-1}(C(+,u)) = S_3(g^{-1}(C(+,1)))$$

In Figure 29 we describe the topology of the eight regions $g^{-1}(R(-,u))$, $g^{-1}(C(-,u))$, $g^{-1}(C(-,1))$, $g^{-1}(R(-,1))$, $g^{-1}(R(+,u))$, $g^{-1}(C(+,u))$, $g^{-1}(C(+,1))$ and $g^{-1}(R(+,1))$.

Now, in an analogous way we show that $f^{-1}(R(-,u))$ is a neighbourhood of $\sigma_+^s(p^-(\pi),\mu)$ on X_1 such that it misses a neighbourhood of $\sigma_-^s(p^-(0),\mu)$ on X_1. The other seven cases are similar. In Figure 30 we give the topology of the eight regions $f^{-1}(R(-,u))$, $f^{-1}(C(-,u))$, $f^{-1}(C(-,1))$, $f^{-1}(R(-,1))$, $f^{-1}(R(+,u))$, $f^{-1}(C(+,u))$, $f^{-1}(C(+,1))$, and $f^{-1}(R(+,1))$.

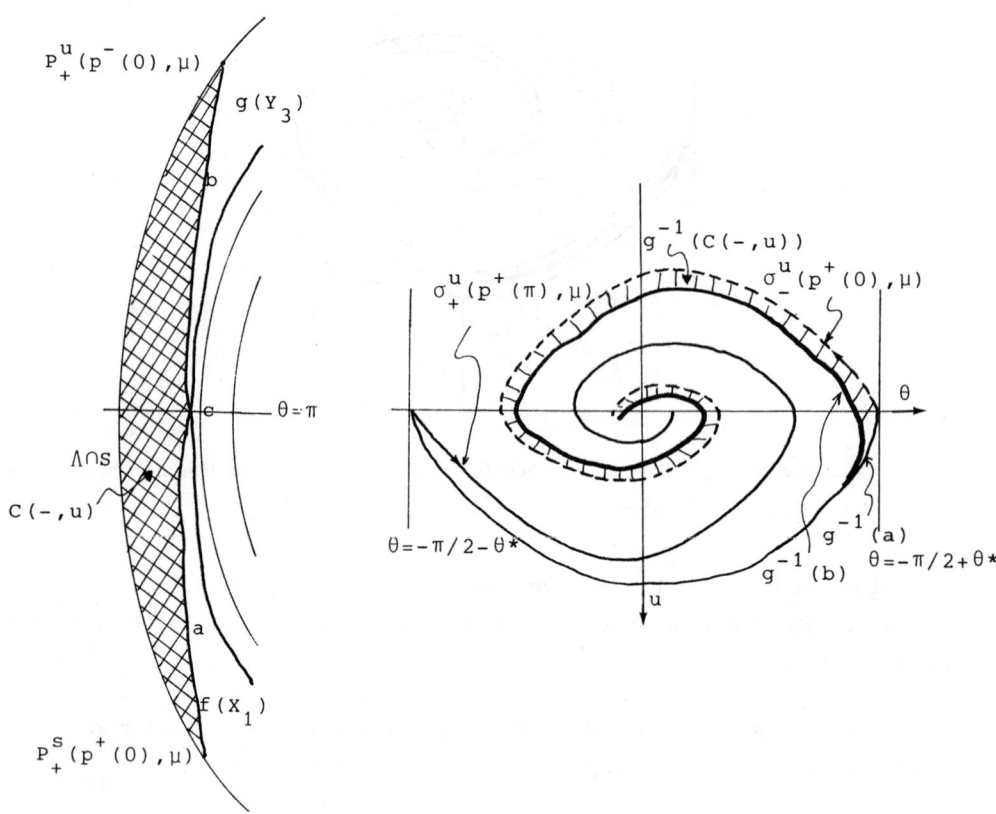

Figure 27. The region C(-,u).   Figure 28. The set $g^{-1}(C(-,u))$.

Here $0 < \theta* = \theta*(\mu,U)$ and U is
the neighbourhood of Theorem 9.

THEOREM 14. For all $\mu > 9/8$ and for all positive integers, $n \geq n_1$, where
$n_1 = n_1(\mu,U)$ and U is the neighbourhood of Theorem 9, the following hold.
(i) The sets (A,n,B) (resp. $(A,n,B]$) are topologically triangular sectors
(resp. curves) on A with a vertex (resp. an endpoint) at the point $P_{\pm}^{S}(\cdot,\mu)$
and its opposite side (resp. the other endpoint) on the opposite side of
the vertex $P_{\pm}^{S}(\cdot,\mu)$ of the sector A, see Figures 31 and 32.
(ii) The sets $h((A,n,B))$ (resp. $[A,n,B))$ are topologically triangular sectors (resp. curves) on B with a vertex (resp. an endpoint) at the point
$P_{\pm}^{u}(\cdot,\mu)$ and its opposite side (resp. the other endpoint) on the opposite
side of the vertex $P_{\pm}^{u}(\cdot,\mu)$ of the sector B, see Figures 33 and 34.
(iii) For each point $P_{\pm}^{S,u}(\cdot,\mu)$ and for each family of triangular sectors
$\{(A,n,B)\}_n$ or $\{h((A,n,b))\}_n$ (resp. curves $\{(A,n,B]\}_n$ or $\{[A,n,B)\}_n$)
which has this point as vertex it accumulates at $\Lambda \cap S$.

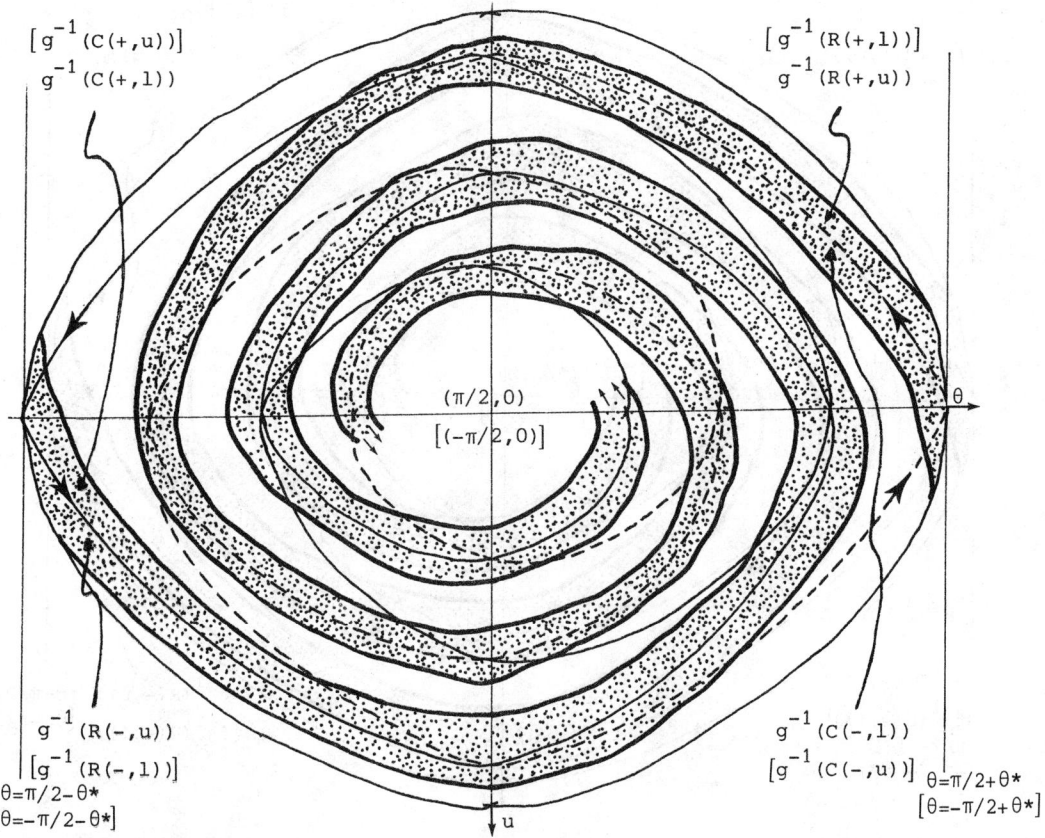

Figure 29. The regions $g^{-1}(R(-,u))$, $g^{-1}(C(-,u))$, $g^{-1}(C(-,1))$, $g^{-1}(R(-,1))$, $g^{-1}(R(+,u))$, $g^{-1}(C(+,u))$, $g^{-1}(C(+,1))$ and $g^{-1}(R(+,1))$.

Proof. We denote by $(g^{-1}(A) \cap f^{-1}(B), n)$ the component of $g^{-1}(A) \cap f^{-1}(B)$ with the number n in Figure 22b. The sets $(g^{-1}(A) \cap f^{-1}(B), n)$ are drawn in Figure 35; this picture is obtained from Figures 22 and the intersection of Figures 29 and 30.

For every A and B such that $g^{-1}(A) \cap f^{-1}(B) \neq \emptyset$ the set $(g^{-1}(A) \cap f^{-1}(B), n)$ is topologically a square contained in $U(\pm\pi/2, \mu)$ such that one side is on $\sigma^u(,\mu)$ and the opposite one is contained on the image under g^{-1} of one of the curves $f(\sigma)$, $f(\Psi)$, $f(\sigma')$ and $f(\Psi')$, see Figures 35, 10 and 36.

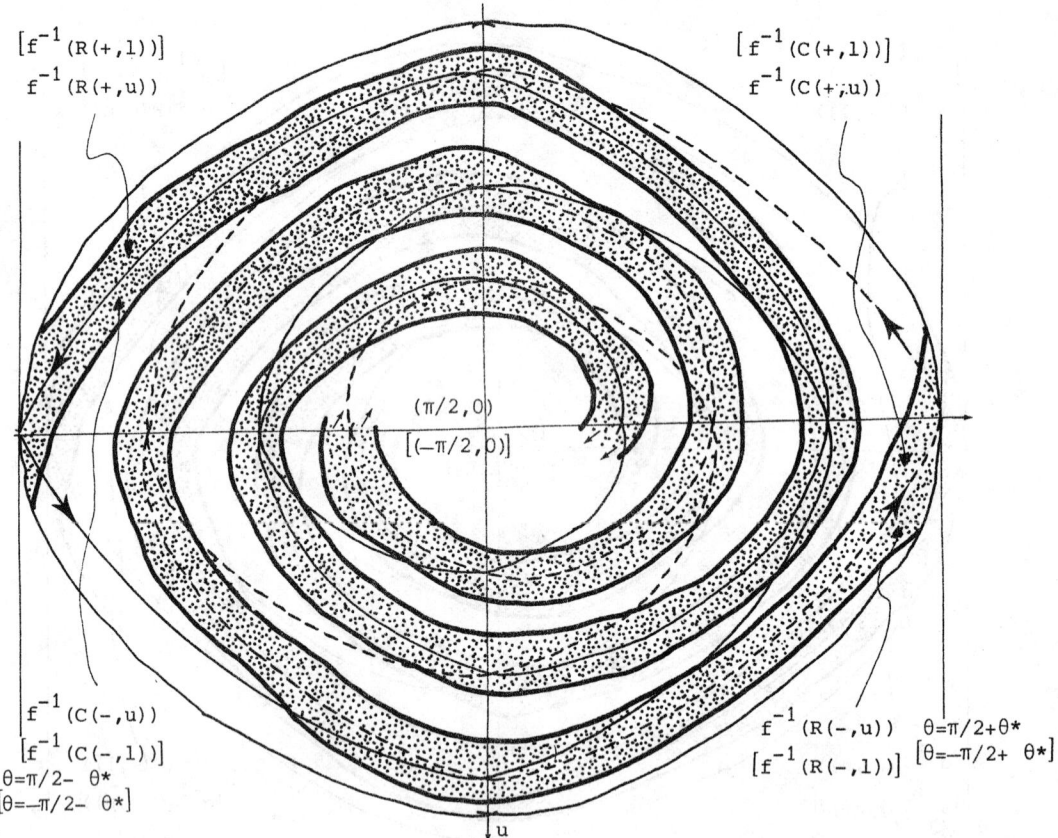

$[f^{-1}(R(+,1))]$
$f^{-1}(R(+,u))$

$[f^{-1}(C(+,1))]$
$f^{-1}(C(+,u))$

$(\pi/2,0)$
$[(-\pi/2,0)]$

$f^{-1}(C(-,u))$
$[f^{-1}(C(-,1))]$
$\theta=\pi/2-\ \theta*$
$[\theta=-\pi/2-\ \theta*]$

$f^{-1}(R(-,u))$ $\theta=\pi/2+\theta*$
$[f^{-1}(R(-,1))]$ $[\theta=-\pi/2+\ \theta*]$

u

Figure 30. The regions $f^{-1}(R(-,u))$, $f^{-1}(C(-,u))$, $f^{-1}(C(-,1))$, $f^{-1}(R(-,1))$, $f^{-1}(R(+,u))$, $f^{-1}(C(+,u))$, $f^{-1}(C(+,1))$, $f^{-1}(R(+,1))$.

From Figure 36 we have that $g(a) = P^s(p,\mu)$ (this equality is topological) and $g(b)$ is contained in $f(\sigma)$, $f(\psi)$, $f(\sigma')$ or $f(\psi')$. The same arguments gives us the topology of $(A,n,B]$. The order of (A,n,B) and $(A,n,B]$ showed in Figures 31 and 32 follows from Figures 22. This completes the proof of (i).

Since $h((A,n,B))=f((g^{-1}(A)\cap f^{-1}(B),n))$, the proof of (ii) is analogous. Part (iii) follows from the fact that the sets $(g^{-1}(A)\cap f^{-1}(B),n)$ accumulate at the points $(\pm\pi/2,0)$, see Figure 35.

Q.E.D.

REMARK 1. The value $n_1(\mu,U)$ only depends on the neighbourhood U of Theorem 9 in the following way : $2n_1(\mu,U)-1 = \min S_{(\pm\pi/2,\ \mu)}(U)$.

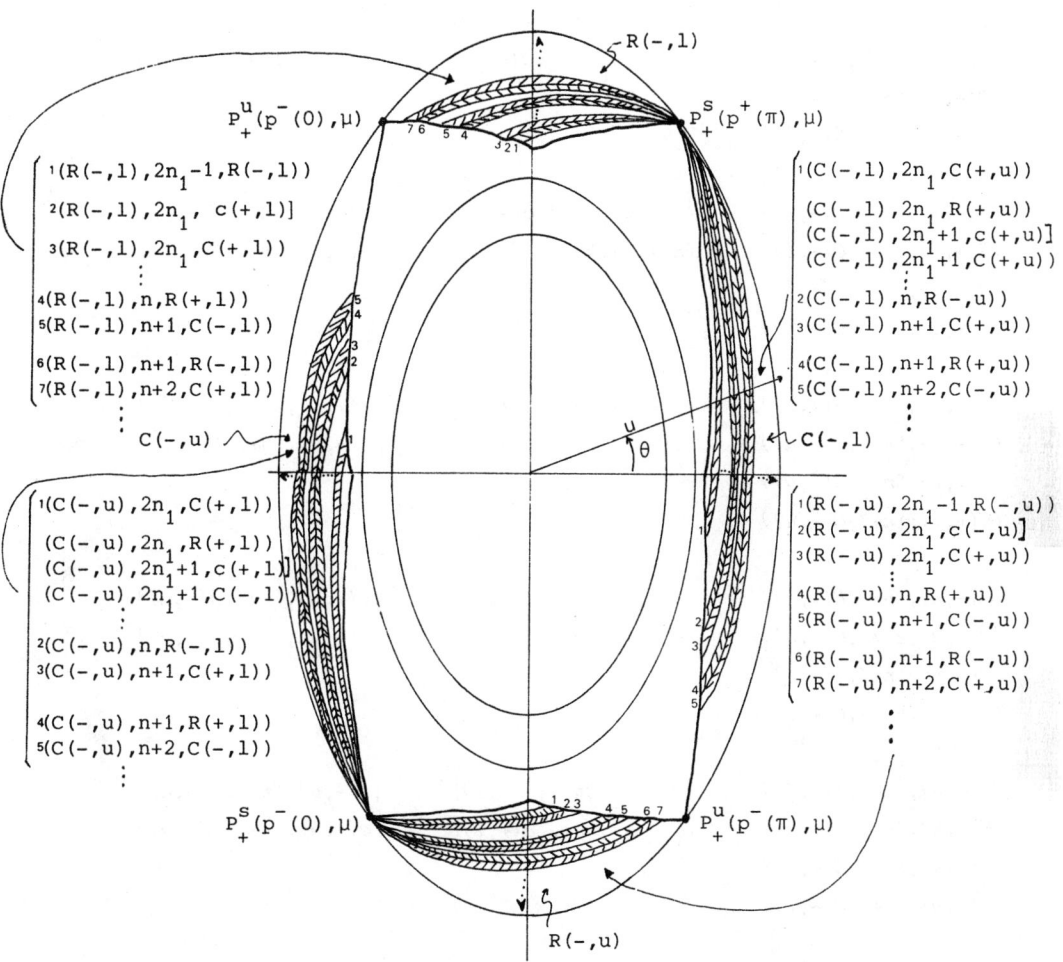

Figure 31. The triangular sectors (A,n,B) and the curves (A,n,B] with u>0.

The following corollary comes from Theorem 14:

COROLLARY 15. For all $\mu > 9/8$ and for all $n \geq n_1$, where $n_1 = n_1(\mu, U)$ and U is the neighbourhood of Theorem 9 the following triads are realizable for the anisotropic Kepler problem.

$(R(+,u), 2n, C(-,u))$	$(R(+,u), 2n+1, c(-,u)]$	$[e(-,u), 2n+1, R(+,u))$
$(R(+,u), 2n-1, R(+,u))$	$(R(+,u), 2n, c(+,u)]$	$[e(+,u), 2n, R(+,u))$
$(R(+,u), 2n+1, C(+,u))$		
$(R(+,u), 2n, R(-,u))$		

$(C(+,u),2n+1,R(+,l))$ $(C(+,u),2n\ \ ,c(+,l)]$ $[e(+,u),2n\ \ ,C(+,u))$

$(C(+,u),2n\ \ ,C(-,l))$ $(C(+,u),2n+1,c(-,l)]$ $[e(-,u),2n+1,C(+,u))$

$(C(+,u),2n\ \ ,R(-,l))$

$(C(+,u),2n+1,C(+,l))$

$(R(+,l),2n\ \ ,R(-,l))$ $(R(+,l),2n+1,c(-,l)]$ $[e(-,l),2n+1,R(+,l))$

$(R(+,l),2n+1,C(+,l))$ $(R(+,l),2n\ \ ,c(+,l)]$ $[e(+,l),2n\ \ ,R(+,l))$

$(R(+,l),2n-1,R(+,l))$

$(R(+,l),2n\ \ ,C(-,l))$

$(C(+,l),2n+1,R(+,u))$ $(C(+,l),2n\ \ ,c(+,u)]$ $[e(+,l),2n\ \ ,C(+,l))$

$(C(+,l),2n\ \ ,C(-,u))$ $(C(+,l),2n+1,c(-,u)]$ $[e(-,l),2n+1,C(+,l))$

$(C(+,l),2n\ \ ,R(-,u))$

$(C(+,l),2n+1,C(+,u))$

$(R(-,u),2n\ \ ,R(+,u))$ $(R(-,u),2n+1,c(+,u)]$ $[e(+,u),2n+1,R(-,u))$

$(R(-,u),2n+1,C(-,u))$ $(R(-,u),2n\ \ ,c(-,u)]$ $[e(-,u),2n\ \ ,R(-,u))$

$(R(-,u),2n-1,R(-,u))$

$(R(-,u),2n\ \ ,C(+,u))$

$(C(-,u),2n+1,R(-,l))$ $(C(-,u),2n\ \ ,c(-,l)]$ $[e(-,u),2n\ \ ,C(-,u))$

$(C(-,u),2n\ \ ,C(+,l))$ $(C(-,u),2n+1,c(+,l)]$ $[e(+,u),2n+1,C(-,u))$

$(C(-,u),2n\ \ ,R(+,l))$

$(C(-,u),2n+1,C(-,l))$

$(R(-,l),2n\ \ ,R(+,l))$ $(R(-,l),2n+1,c(+,l)]$ $[e(+,l),2n+1,R(-,l))$

$(R(-,l),2n+1,C(-,l))$ $(R(-,l),2n\ \ ,c(-,l)]$ $[e(-,l),2n\ \ ,R(-,l))$

$(R(-,l),2n-1,R(-,l))$

$(R(-,l),2n\ \ ,C(+,l))$

$(C(-,l),2n+1,R(-,u))$ $(C(-,l),2n\ \ ,c(-,u)]$ $[e(-,l),2n\ \ ,C(-,l))$

$(C(-,l),2n\ \ ,C(+,u))$ $(C(-,l),2n+1,c(+,u)]$ $[e(+,l),2n+1,C(-,l))$

$(C(-,l),2n\ \ ,R(+,u))$

$(C(-,l),2n+1,C(-,u))$

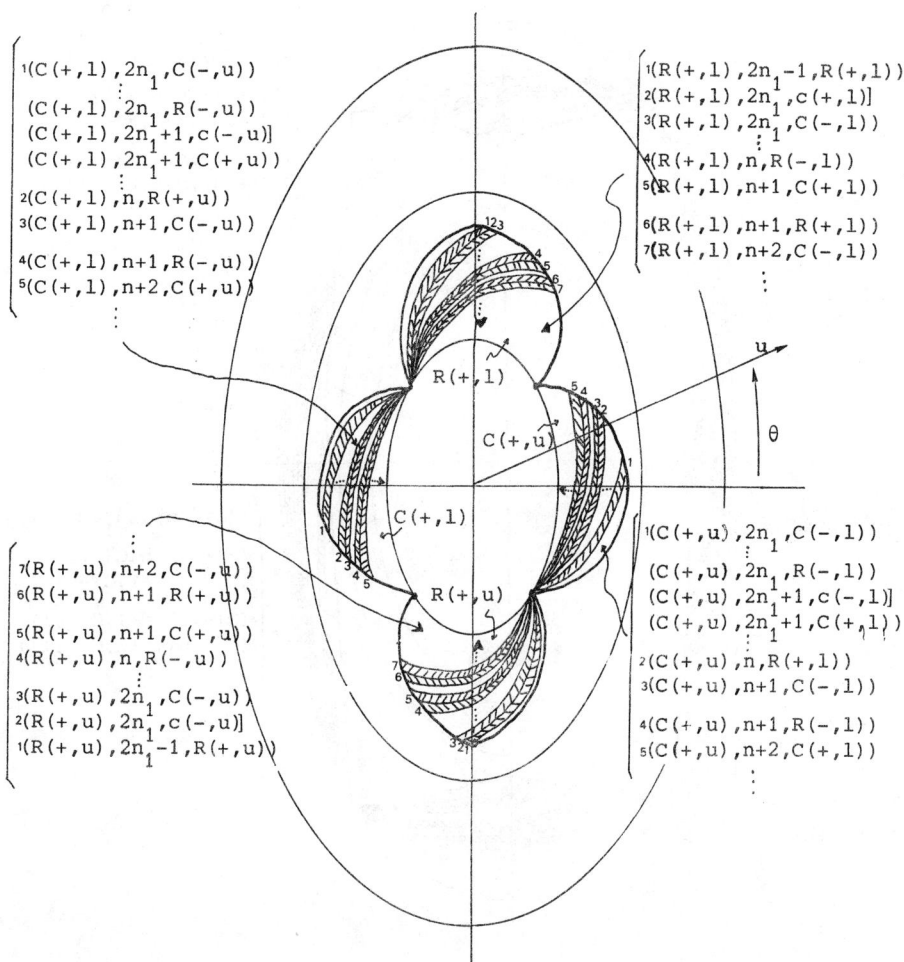

$^1(C(+,1),2n_1,C(-,u))$
$(C(+,1),2n_1,R(-,u))$
$(C(+,1),2n_1+1,c(-,u)]$
$(C(+,1),2n_1+1,C(+,u))$
$^2(C(+,1),n,R(+,u))$
$^3(C(+,1),n+1,C(-,u))$
$^4(C(+,1),n+1,R(-,u))$
$^5(C(+,1),n+2,C(+,u))$

$^1(R(+,1),2n_1-1,R(+,1))$
$^2(R(+,1),2n_1,c(+,1)]$
$^3(R(+,1),2n_1,C(-,1))$
$^4(R(+,1),n,R(-,1))$
$^5(R(+,1),n+1,C(+,1))$
$^6(R(+,1),n+1,R(+,1))$
$^7(R(+,1),n+2,C(-,1))$

$^7(R(+,u),n+2,C(-,u))$
$^6(R(+,u),n+1,R(+,u))$
$^5(R(+,u),n+1,C(+,u))$
$^4(R(+,u),n,R(-,u))$
$^3(R(+,u),2n_1,C(-,u))$
$^2(R(+,u),2n_1,c(-,u)]$
$^1(R(+,u),2n_1-1,R(+,u))$

$^1(C(+,u),2n_1,C(-,1))$
$(C(+,u),2n_1,R(-,1))$
$(C(+,u),2n_1+1,c(-,1)]$
$(C(+,u),2n_1+1,C(+,1))$
$^2(C(+,u),n,R(+,1))$
$^3(C(+,u),n+1,C(-,1))$
$^4(C(+,u),n+1,R(-,1))$
$^5(C(+,u),n+2,C(+,1))$

$R(+,1)$
$C(+,u)$
$C(+,1)$
$R(+,u)$

u

θ

Figure 32. The triangular sectors (A,n,B) and the curves $(A,n,B]$ with $u<0$.

The triad $[A,n,B]$ where $A\in\{e(+,u),e(-,u),e(+,1),e(-,1)\}$ and $B\in\{c(+,u),c(-,u),c(+,1),c(-,1)\}$ is formed by the orbits which start in ejection and end in collision having to meet the q_2-axis n times. Since the points of $[A,n,B]$ are in $\sigma^u(\ ,\mu)\cap\sigma^s(\ ,\mu)$, neither f nor q are defined on them. From Lemma 7 and Figure 22b we have,

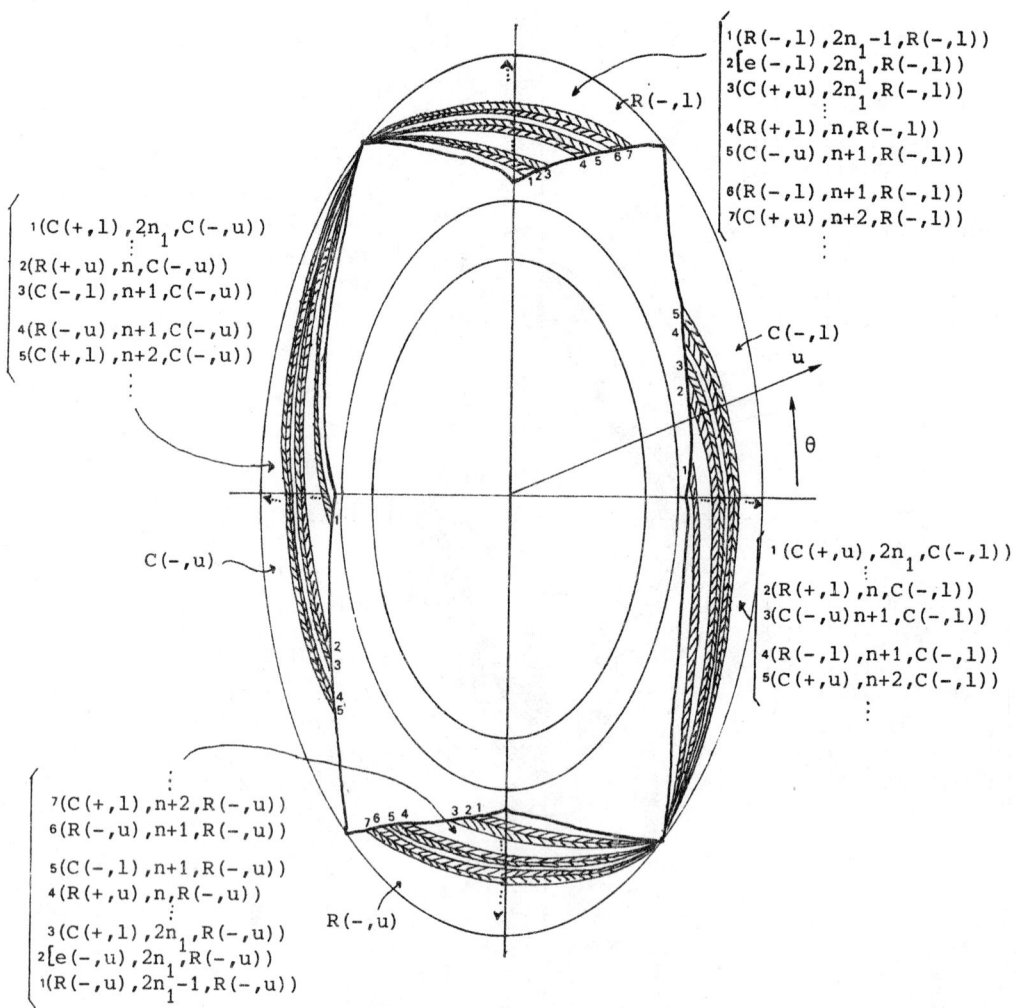

Figure 33. The triangular sectors h((A,n,B)) and the curves [A,n,B) with u>0.

COROLLARY 16. For all μ>9/8 and for all n ≥ n₁+1, where n₁=n₁(μ,U) and U is the neighbourhood of Theorem 9, the following triads are realizable for the anisotropic Kepler problem:

$[e(+,u),2n-1,c(+,u)]$

$[e(+,u),2n\ \ ,c(-,u)]$

$[e(-,u),2n\ \ ,c(+,u)]$

$[e(-,u),2n-1,c(-,u)]$

$[e(+,l),2n-1,c(+,l)]$

$[e(+,l),2n\ \ ,c(-,l)]$

$[e(-,l),2n\ \ ,c(+,l)]$

$[e(-,l),2n-1,c(-,l)]$

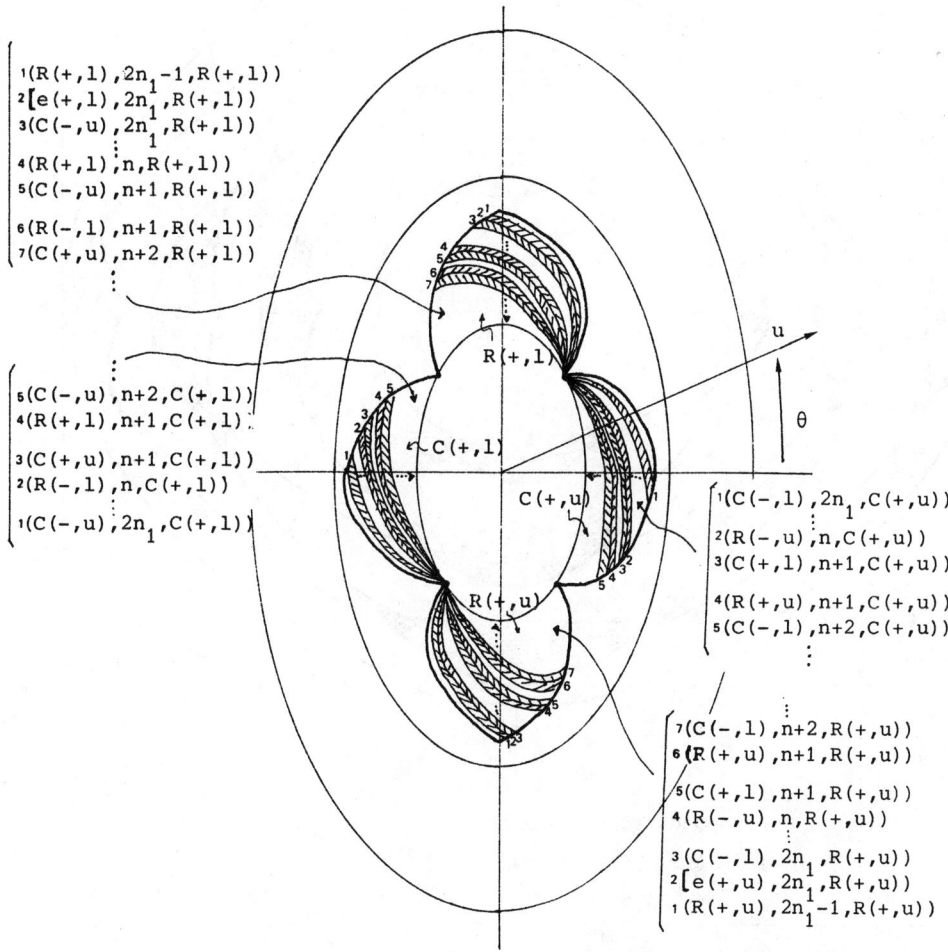

$1(R(+,1),2n_1-1,R(+,1))$
$2[e(+,1),2n_1,R(+,1))$
$3(C(-,u),2n_1,R(+,1))$
$\vdots$
$4(R(+,1),n,R(+,1))$
$5(C(-,u),n+1,R(+,1))$

$6(R(-,1),n+1,R(+,1))$
$7(C(+,u),n+2,R(+,1))$
$\vdots$

$\vdots$
$5(C(-,u),n+2,C(+,1))$
$4(R(+,1),n+1,C(+,1))$
$3(C(+,u),n+1,C(+,1))$
$2(R(-,1),n,C(+,1))$
$1(C(-,u),2n_1,C(+,1))$

$1(C(-,1),2n_1,C(+,u))$
$2(R(-,u),n,C(+,u))$
$3(C(+,1),n+1,C(+,u))$
$4(R(+,u),n+1,C(+,u))$
$5(C(-,1),n+2,C(+,u))$
$\vdots$

$7(C(-,1),n+2,R(+,u))$
$6(R(+,u),n+1,R(+,u))$
$5(C(+,1),n+1,R(+,u))$
$4(R(-,u),n,R(+,u))$
$\vdots$
$3(C(-,1),2n_1,R(+,u))$
$2[e(+,u),2n_1,R(+,u))$
$1(R(+,u),2n_1-1,R(+,u))$

Figure 34. The triangular sectors $h((A,n,B))$ and the curves $[A,n,B)$ with $u< 0$.

$a \subset \sigma^u(\ ,\mu)$

$(g^{-1}(A) \cap f^{-1}(B),n)$

$b \subset g^{-1}(f(\sigma) \cup f(\psi) \cup f(\sigma') \cup f(\psi'))$

Figure 36. Topological description of the set $(g^{-1}(A) \cap f^{-1}(B),n)$.

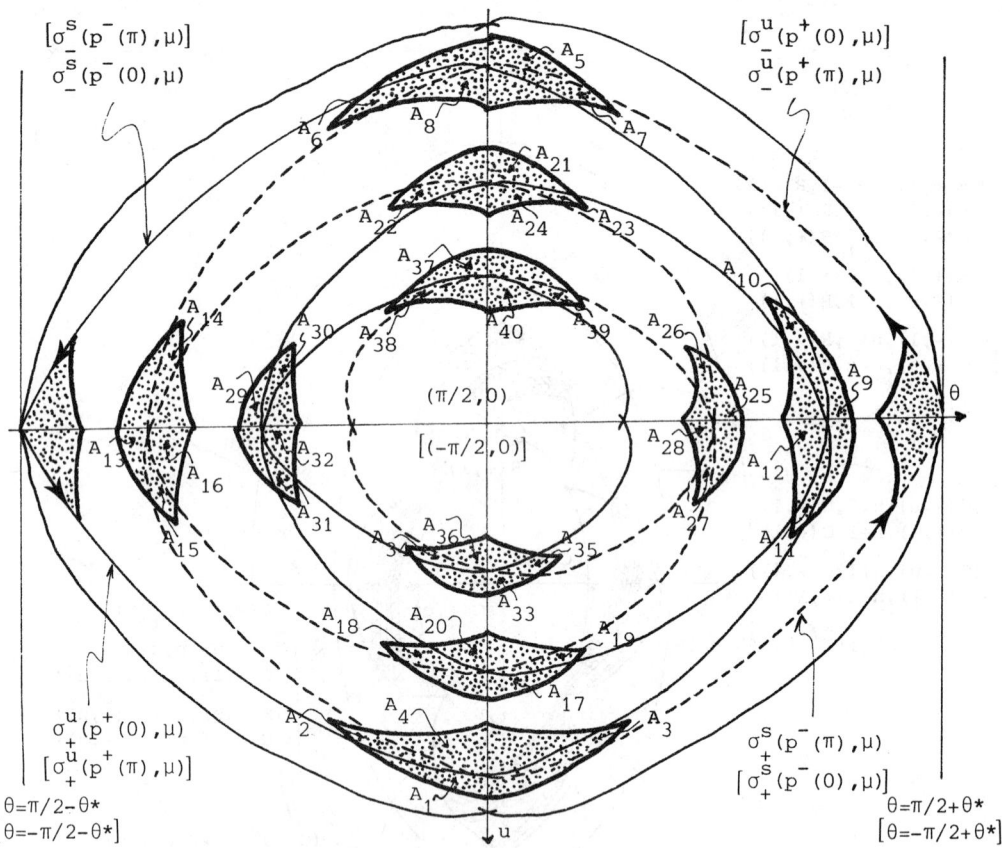

The region	Corresponds to the set
A_1	$(g^{-1}(R(+,u) \cap f^{-1}(R(+,u), 2n_1-1)$
	$[(g^{-1}(R(-,1) \cap f^{-1}(R(+,1), 2n_1-1)]$
A_2	$(g^{-1}(C(-,1) \cap f^{-1}(R(+,u), 2n_1)$
	$[(g^{-1}(C(-,u) \cap f^{-1}(R(+,1), 2n_1)]$
A_3	$(g^{-1}(R(+,u) \cap f^{-1}(C(-,u), 2n_1)$
	$[(g^{-1}(R(+,1) \cap f^{-1}(C(-,1), 2n_1)]$
A_4	$(g^{-1}(C(-,1) \cap f^{-1}(C(-,u), 2n_1+1)$
	$[(g^{-1}(C(-,u) \cap f^{-1}(C(-,1), 2n_1+1)]$
A_5	$(g^{-1}(R(-,u) \cap f^{-1}(R(-,u), 2n_1-1)$
	$[(g^{-1}(R(-,1) \cap f^{-1}(R(-,1), 2n_1-1)]$
A_6	$(g^{-1}(R(-,u) \cap f^{-1}(C(+,u), 2n_1)$
	$[(g^{-1}(R(-,1) \cap f^{-1}(C(+,1), 2n_1)]$
A_7	$(g^{-1}(C(+,1) \cap f^{-1}(R(-,u), 2n_1)$
	$[(g^{-1}(C(+,u) \cap f^{-1}(R(-,1), 2n_1)]$

The region	Corresponds to the set
A_8	$(g^{-1}(C(+,1)) \cap f^{-1}(C(+,u)),\ 2n_1+1)$ $[(g^{-1}(C(+,u)) \cap f^{-1}(C(+,1)),\ 2n_1+1)]$
A_9	$(g^{-1}(R(-,u)) \cap f^{-1}(R(+,u)),\ 2n_1)$ $[(g^{-1}(R(-,1)) \cap f^{-1}(R(+,1)),\ 2n_1)]$
A_{10}	$(g^{-1}(R(-,u)) \cap f^{-1}(C(-,u)),\ 2n_1+1)$ $[(g^{-1}(R(-,1)) \cap f^{-1}(C(-,1)),\ 2n_1+1)]$
A_{11}	$(g^{-1}(C(+,1)) \cap f^{-1}(R(+,u)),\ 2n_1+1)$ $[(g^{-1}(C(+,u)) \cap f^{-1}(R(+,1)),\ 2n_1+1)]$
A_{12}	$(g^{-1}(C(+,1)) \cap f^{-1}(C(-,u)),\ 2n_1+2)$ $[(g^{-1}(C(+,u)) \cap f^{-1}(C(-,1)),\ 2n_1+2)]$
A_{13}	$(g^{-1}(R(+,u)) \cap f^{-1}(R(-,u)),\ 2n_1)$ $[(g^{-1}(R(+,1)) \cap f^{-1}(R(-,1)),\ 2n_1)]$
A_{14}	$(g^{-1}(R(+,u)) \cap f^{-1}(C(+,u)),\ 2n_1+1)$ $[(g^{-1}(R(+,1)) \cap f^{-1}(C(+,1)),\ 2n_1+1)]$
A_{15}	$(g^{-1}(C(-,1)) \cap f^{-1}(R(-,u)),\ 2n_1+1)$ $[(g^{-1}(C(-,u)) \cap f^{-1}(R(-,1)),\ 2n_1+1)]$
A_{16}	$(g^{-1}(C(-,1)) \cap f^{-1}(C(+,u)),\ 2n_1+2)$ $[(g^{-1}(C(-,u)) \cap f^{-1}(C(+,1)),\ 2n_1+2)]$
A_{17}	$(g^{-1}(R(-,u)) \cap f^{-1}(R(-,u)),\ 2n_1+1)$ $[(g^{-1}(R(-,1)) \cap f^{-1}(R(-,1)),\ 2n_1+1)]$
A_{18}	$(g^{-1}(C(+,1)) \cap f^{-1}(R(-,u)),\ 2n_1+2)$ $[(g^{-1}(C(+,u)) \cap f^{-1}(R(-,1)),\ 2n_1+2)]$
A_{19}	$(g^{-1}(R(-,u)) \cap f^{-1}(C(+,u)),\ 2n_1+2)$ $[(g^{-1}(R(-,1)) \cap f^{-1}(C(+,1)),\ 2n_1+2)]$
A_{20}	$(g^{-1}(C(+,1)) \cap f^{-1}(C(+,u)),\ 2n_1+3)$ $[(g^{-1}(C(+,u)) \cap f^{-1}(C(+,1)),\ 2n_1+3)]$
A_{21}	$(g^{-1}(R(+,u)) \cap f^{-1}(R(+,u)),\ 2n_1+1)$ $[(g^{-1}(R(+,u)) \cap f^{-1}(R(+,u)),\ 2n_1+1)]$
A_{22}	$(g^{-1}(R(+,u)) \cap f^{-1}(C(-,u)),\ 2n_1+2)$ $[(g^{-1}(R(+,1)) \cap f^{-1}(C(-,1)),\ 2n_1+2)]$
A_{23}	$(g^{-1}(C(-,1)) \cap f^{-1}(R(+,u)),\ 2n_1+2)$ $[(g^{-1}(C(-,u)) \cap f^{-1}(R(+,1)),\ 2n_1+2)]$

A_{24}	$(g^{-1}(C(-,1)) \cap f^{-1}(C(-,u)), \ 2n_1+3)$ $[(g^{-1}(C(-,u)) \cap f^{-1}(C(-,1)), \ 2n_1+3)]$
A_{25}	$(g^{-1}(R(+,u)) \cap f^{-1}(R(-,u)), \ 2n_1+2)$ $[(g^{-1}(R(+,1)) \cap f^{-1}(R(-,1)), \ 2n_1+2)]$
A_{26}	$(g^{-1}(R(+,u)) \cap f^{-1}(C(+,u)), \ 2n_1+3)$ $[(g^{-1}(R(+,1)) \cap f^{-1}(C(+,1)), \ 2n_1+3)]$
A_{27}	$(g^{-1}(C(-,1)) \cap f^{-1}(R(-,u)), \ 2n_1+3)$ $[(g^{-1}(C(-,u)) \cap f^{-1}(R(-,1)), \ 2n_1+3)]$
A_{28}	$(g^{-1}(C(-,1)) \cap f^{-1}(C(+,u)), \ 2n_1+4)$ $[(g^{-1}(C(-,u)) \cap f^{-1}(C(+,1)), \ 2n_1+4)]$
A_{29}	$(g^{-1}(R(-,u)) \cap f^{-1}(R(+,u)), \ 2n_1+2)$ $[(g^{-1}(R(-,1)) \cap f^{-1}(R(+,1)), \ 2n_1+2)]$
A_{30}	$(g^{-1}(C(+,1)) \cap f^{-1}(R(+,u)), \ 2n_1+3)$ $[(g^{-1}(C(+,u)) \cap f^{-1}(R(+,1)), \ 2n_1+3)]$
A_{31}	$(g^{-1}(R(-,u)) \cap f^{-1}(C(-,u)), \ 2n_1+3)$ $[(g^{-1}(R(-,1)) \cap f^{-1}(R(-.1)), \ 2n_1+3)]$
A_{32}	$(g^{-1}(C(+,1)) \cap f^{-1}(C(-,u)), \ 2n_1+4)$ $[(g^{-1}(C(+,u)) \cap f^{-1}(C(-,1)), \ 2n_1+4)]$
A_{33}	$(g^{-1}(R(+,u)) \cap f^{-1}(R(+,u)), \ 2n_1+3)$ $[(g^{-1}(R(+,1)) \cap f^{-1}(R(+,1)), \ 2n_1+3)]$
A_{34}	$(g^{-1}(C(-,1)) \cap f^{-1}(R(+,u)), \ 2n_1+4)$ $[(g^{-1}(C(-,u)) \cap f^{-1}(R(+,1)), \ 2n_1+4)]$
A_{35}	$(g^{-1}(R(+,u)) \cap f^{-1}(C(-,u)), \ 2n_1+4)$ $[(g^{-1}(R(+,1)) \cap f^{-1}(C(-,1)), \ 2n_1+4)]$
A_{36}	$(g^{-1}(R(-,u)) \cap f^{-1}(R(-,u)), \ 2n_1+5)$ $[(g^{-1}(R(-,1)) \cap f^{-1}(R(-,1)), \ 2n_1+5)]$
A_{37}	$(g^{-1}(R(-,u)) \cap f^{-1}(R(-,u)), \ 2n_1+3)$ $[(g^{-1}(R(-,1)) \cap f^{-1}(R(-,1)), \ 2n_1+3)]$
A_{38}	$(g^{-1}(R(-,u)) \cap f^{-1}(C(+,u)), \ 2n_1+4)$ $[(g^{-1}(R(-,1)) \cap f^{-1}(C(+,1)), \ 2n_1+4)]$
A_{39}	$(g^{-1}(C(+,1)) \cap f^{-1}(R(-,u)), \ 2n_1+4)$ $[(g^{-1}(C(+,u)) \cap f^{-1}(R(-,1)), \ 2n_1+4)]$
A_{40}	$(g^{-1}(C(+,1)) \cap f^{-1}(C(+,u)), \ 2n_1+5)$ $[(g^{-1}(C(+,u)) \cap f^{-1}(C(+,1)), \ 2n_1+5)]$

$\vdots$

Figure 35. The sets $(g^{-1}(A) \cap f^{-1}(B), \ n)$

(IV.7) A subshift as subsystem of h .

We consider the set of sequences $\{T_n\}$, where n belongs to the integers, such that T_n is a triad of Corollaries 15 and 16, and $T_{n+1}=(A,r,B)$ or $T_{n+1}=(A,r,B]$ can follow to $T_n=(A',r',B')$ or $T_n=[A',r',B')$ if and only if B'=A (we call these two triads compatible).

We shall be interested in the next four types of sequences $\{T_n\}$:

(a) For all $n\in \mathbb{Z}$ there exists T_n of type (A,m,B). We denote this type of sequences by,

$$(\ \ldots\ ,\ T_{-2},T_{-1},T_0,T_1,T_2,\ \ldots\ )$$

(b) Let k be a negative integer. Then for all n>k+1 there exists T_n of type (A,m,B) but T_k does not exists and T_{k+1} is of type [A,m,B). We write this type of sequences by,

$$[\ \ T_{k+1},T_{k+2},\ \ldots\ )$$

(c) Let l be a positive integer. Then for all n<l-1 there exists T_n of type (A,m,B) but T_l does not exists and T_{l-1} is of type (A,m,B], and we write,

$$(\ \ldots,\ T_{l-2},T_{l-1}\ \ ]$$

(d) Let k and l be integers such that k<0 and l>0. Then for all k+1<n<l-1 there exists T_n of type (A,m,B) but T_l and T_k do not exist, T_{k+1} is of type [A,m,B) and T_{l-1} is of type (A,m,B], and we write,

$$[\ \ T_{k+1},T_{k+2},\ \ldots\ ,\ T_{l-2},T_{l-1}\ \ ]$$

To each sequence of the above types we shall associate a solution of the anisotropic Kepler problem for $\mu > 9/8$. Sequences of type (a) will correspond to orbits without ejection and without collision. Sequences (b) will correspond to orbits with ejection and without collision. Sequences (c) will be associated to orbits without ejection and with collision and, finally, sequences (d) will be for ejection-collision orbits.

Let $p(\tau) = (r(\tau), v(\tau), \theta(\tau), u(\tau))$ be a solution of the anisotropic
Kepler problem such that $p(0)$ belongs to the domain of definition of h, and
let $\{T_n\}$ be a sequence of type (a). We say that $p(\tau)$ <u>realizes</u> $\{T_n\}$ if
$h^n(p(0)) \in T_n$ for all $n \in \mathbb{Z}$. In the same way we define the realization of a
sequence of type (b), (c) or (d) by a solution of the anisotropic Kepler
problem.

Figures 37 show some sequences and solutions which realize them.

The basic result of this section is:

<u>THEOREM 17</u>. *For all $\mu > 9/8$ and every sequence $\{T_n\}$ of type (a), (b), (c) or
(d) there exists a solution of the anisotropic Kepler problem which realizes
it.*

<u>Proof</u>. Let T_n and T_{n+1} be two compatible triads. We shall describe the topo-
logy of the set,

$$T_n T_{n+1} = \{ p \in T_n : h(p) \in T_{n+1} \} = T_n \cap h^{-1}(T_{n+1})$$

If $T_n = (C, m, A)$ and $T_{n+1} = (A, m', B)$, from Theorem 14 we have that T_{n+1}
meets $h(T_n)$ as it is shown in Figure 38 (see Figures 31,32,33 and 34).

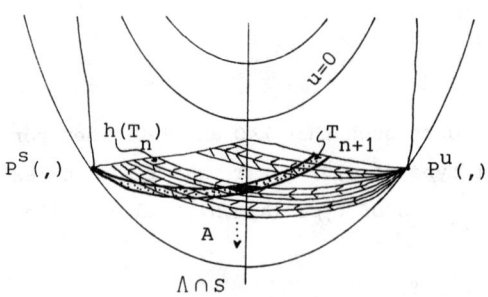

<u>Figure 38</u>. The set $h(T_n) \cap T_{n+1}$.

From Figure 30 it follows that $f^{-1}(T_{n+1}) \subset f^{-1}(A) \subset X_i$, where $i \in \{1,2,3,4\}$,
is a strip which spirals to $\gamma_h(\overset{+}{-}\pi/2) \cap S$, see Figure 39. In particular,

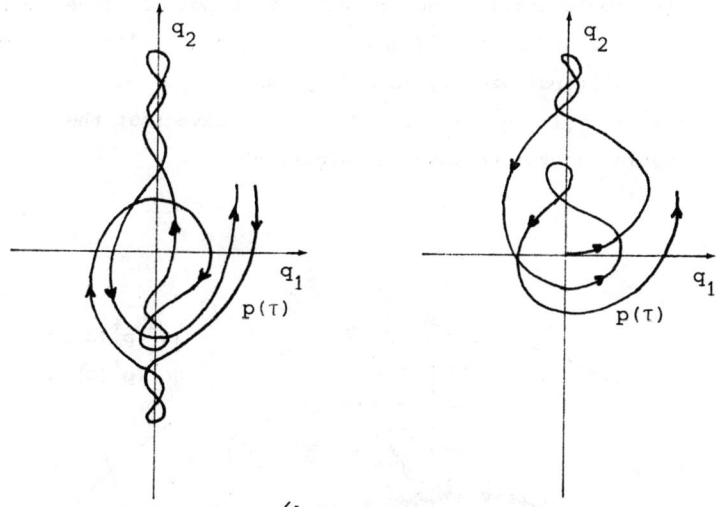

$$\big([e(-,u),6,R(-,u)),(R(-,u),3,R(-,u)),\ldots\big)$$
$$(\ldots,(C(+,u),5,R(+,1)),(R(+,1),4,C(-,1)),(C(-,1),7,R(-,u)),\ldots)$$

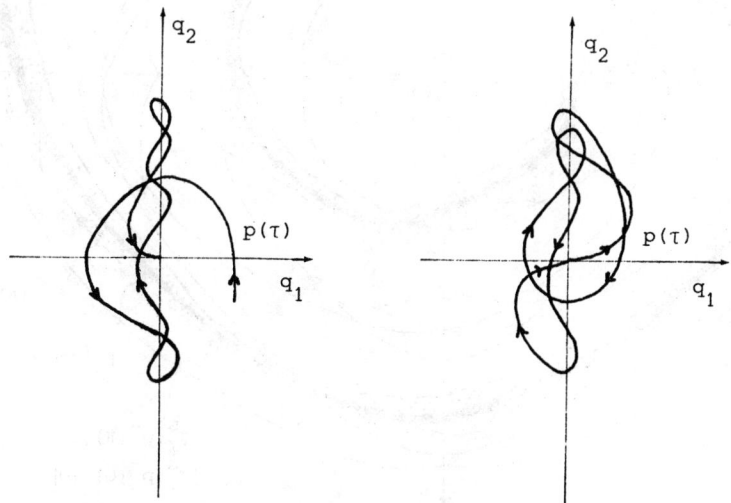

$$(\ldots,(R(-,1),4,C(+,1)),(C(+,1),7,c(-,u)])$$
$$\big([e(-,u),3,R(+,u)),(R(+,u),4,C(\,,u)),(C(\,,u),3,c(+,1)]\big)$$

Figure 37. Solutions p(τ) which realize the indicated sequences.

$g^{-1}(T_n) \cap f^{-1}(T_{n+1})$ is topologically a square with two opposite sides such that one is on $\sigma^u(\ ,\mu)$ and the other on $g^{-1}(f(\sigma) \cup f(\psi) \cup f(\sigma') \cup f(\psi'))$ (see Figure 35). Since $g(\sigma^u(\ ,\mu))$ topologically is $P^s(\ ,\mu)$ and

$T_n \cap h^{-1}(T_{n+1}) = T_n \cap g \circ f^{-1}(T_{n+1}) = g(g^{-1}(T_n) \cap f^{-1}(T_{n+1}))$, we have that the set $T_n T_{n+1}$ is the triangular sector shadowed in Figure 40.

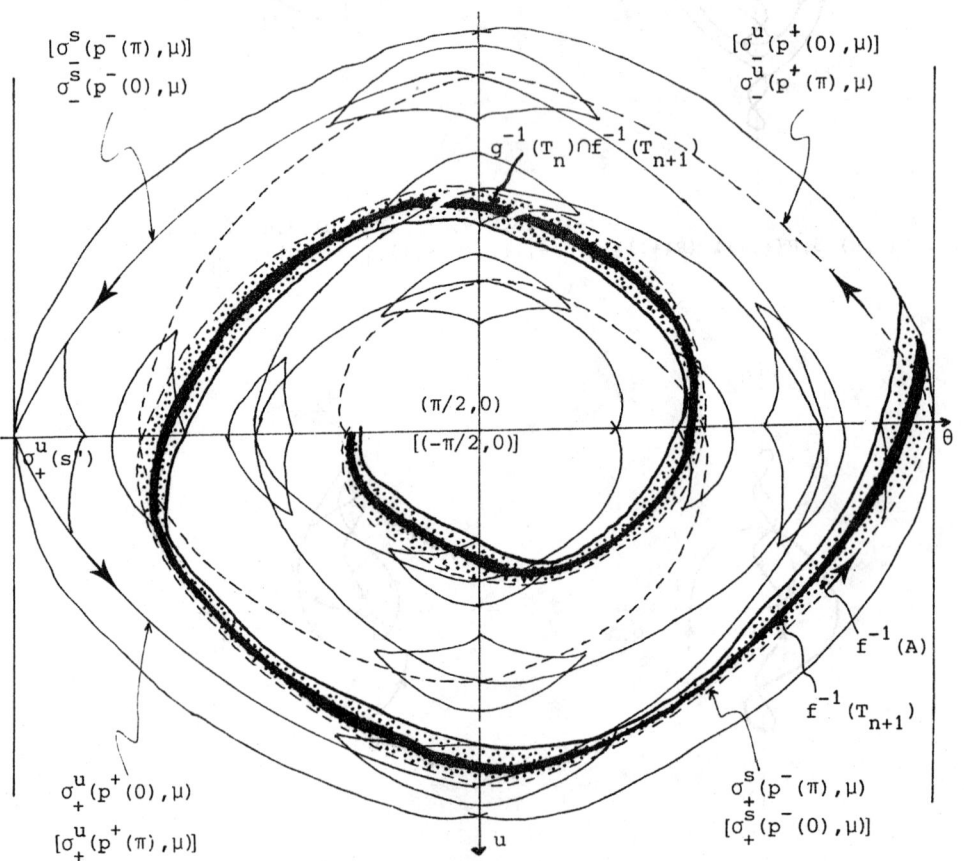

<u>Figure 39</u>. The set $g^{-1}(T_n) \cap f^{-1}(T_{n+1})$.

Forthwith, we prove the theorem for the sequences $\{T_n\}_{n \in \mathbb{Z}}$ of type (a).

We claim that the set ,

$$Z=\{p\in T_0: h^n(p)\in T_n \text{ for all } n\geqslant 1\} = T_0\cap h^{-1}(T_1)\cap h^{-2}(T_2)\cap \dots$$

is compact, non-empty and contains an arc joining $P^s(\ ,\mu)$ to the opposite side in the triangular sector A if $T_0=(A,n,B)$. The proof of this claim also show us that the set,

$$Y=\{p\in T_0: h^n(p)\in T_n \text{ for all } n\leqslant -1\} = T_0\cap h(T_{-1})\cap h^2(T_{-2})\cap \dots$$

is compact, non-empty and contains an arc joining $P^u(\ ,\mu)$ to the opposite side in A. By using both results we obtain for the sequence $\{T_n\}_{n\in\mathbb{Z}}$ of type (a) the existence of at least one point $p\in Z\cap Y$ which, by construction, realizes $\{T_n\}_{n\in\mathbb{Z}}$, see Figure 41.

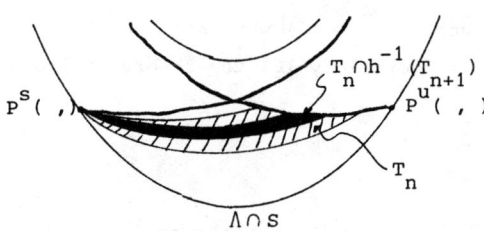

Figure 40. The set $T_n\cap h^{-1}(T_{n+1})$.

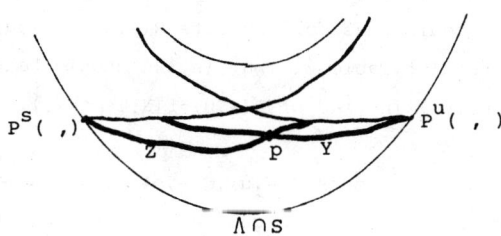

Figure 41. A point p which realizes the sequence $\{T_n\}_{n\in\mathbb{Z}}$.

Now, we shall prove the claim. For all $m>0$ we consider the set, $Z_m = \{p_o \in T_o : h^n(p) \in T_n\}$ for $1 \leqslant n \leqslant m$. We shall show that Z_m is compact, non-empty and contains an arc going from $P^s(,\mu)$ to the opposite side in A. Then, since $Z = \bigcap_{m \in \mathbb{N}} Z_m$ and $Z_m \supset Z_{m+1}$ the claim follows.

We use induction with respect to m. For $m=1$ we have proved it in Figure 40; on the other hand, we have $Z_{m+1} = T_o \cap h^{-1}(Z_m')$ where, $Z_m' = \{p \in T_1 : h^n(p) \in T_{n+1}$ for $1 \leqslant n \leqslant m-1\}$. By hypotheses of induction, Z_m' satisfies the required condition and then similar arguments used in order to obtain Figure 40, prove that Z_{m+1} is compact, non-empty and contains an arc going from $P^s(,\mu)$ to the opposite side in A.

Let $[T_{k+1}, T_{k+2}, \ldots)$ be a sequence of type (b). Since $T_{k+1} = [B,n,A)$ is an arc going from $P^u(,\mu)$ to the opposite side in A (see Theorem 14), $T_{k+2} \cap h(T_{k+1})$ is another arc satisfying the same conditions. So the same is true for $Y = T_o \cap h(T_{-1}) \cap h^2(T_{-2}) \cap \ldots \cap h^{-k-1}(T_{k+1})$. On the other hand, Z is the same as for a sequence of type (a). Then the proof of the theorem follows in an analogous way for a sequence of type (b).

For sequences of type (c) and (d) when $k=-1$ and $l=1$ (i.e., a sequence $[T_o]$ where T_o is a triad of Corollary 16) the theorem follows from Figure 22b.

$\qquad\qquad\qquad\qquad\qquad\qquad$ Q.E.D.

The triads $T_n = \{A,m,B\}$, where $\{$ denotes (or [and $\}$ denotes) or], of the sequences of Theorem 17 are such that m is greater than or equal to either $2n_1-1$ or $2n_1$, or $2n_1+1$ according to Corollaries 15 and 16, where $n_1 = n_1(\mu,U)$.

In order to decrease the value of $n_1(\mu,U)$, for instance $n_1(\mu)=1$, and to improve Theorem 17 we should consider the sets $R(-,u)$, $C(-,u)$, $\ldots$, $R(+,l)$ as in Table 1 instead of Table 2. That is, we should take $R(-,u) = f(U_+^s(p^-(\pi),\mu)) \cap g(U_+^u(p^+(0),\mu))$ instead of $R(-,u) = f(X_1) \cap g(Y_1)$.

Now, if we study one of the sets $R(-,u), C(-,u), \ldots, R(+,l)$ given in Table 1, we obtain a picture similar to Figure 42. In this picture the set $R(-,u)$ is drawn if the neighbourhoods $U_+^s(p^-(\pi),\mu)$ and $U_+^u(p^+(0),\mu)$ are chosen containing the spiral strip given in Figure 26. This choosing of the neighbourhoods $U_{+,-}^{s,u}(,\mu)$ and $V_{+,-}^{s,u}(,\mu)$ is a key point in order to improve Theorem 17

The curves a and b of Figure 42 play the same role as the curves a,b of Figure 25 .Then the set $g^{-1}(R(-,u))$ looks like Figure 26 but now $\theta^*=\pi/2$ if $9/8 \leqslant \mu \leqslant 4$ and $\pi/2-\theta^* =\theta(\sigma_+^u(s''))$ if $\mu>4$ (see (IV.1)).

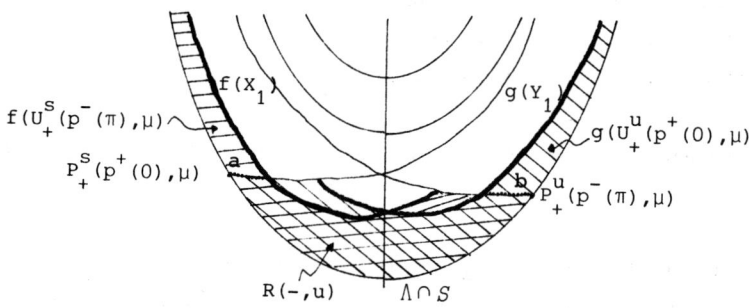

Figure 42. The region $R(-,u)$ given by Table 1 with a convenient election of $U_+^s(p^-(\pi),\mu)$ and $U_+^u(p^+(0),\mu)$

By using the same arguments we can obtain Theorem 14' and Corollaries 15' and 16' similar to Theorem 14 and Corollaries 15 and 16 but with $n_1=1$ if $9/8<\mu<\mu_c$ and $n_1=n_0(\mu)$ given by Theorem 13 if $\mu \geqslant \mu_c$.

Now, we consider the set of sequences $\{T_n'\}$ of types (a),(b),(c) and (d) where the triads T_n' belong to Corollaries 15' and 16'. Then, in an analogous way to Theorem 17 we obtain the next result.

THEOREM 17'. _For all $\mu>9/8$ and every sequence $\{T_n'\}$ of type (a),(b),(c) and (d) there exists a solution of the anisotropic Kepler problem which realizes it._

Let A be the set $\mathbb{N} \cup \{\infty\}$ where we have the usual order extended by $a<\infty$ for all $a\in\mathbb{N}$. Let S be the set of sequences of elements belonging to A of the types:

(a) $(\ldots, a_{-2},a_{-1},a_0,a_1,a_2, \ldots)$ with $a_n \neq \infty$ for all $n\in\mathbb{Z}$,

(b) $[\infty,a_{k+1},a_{k+2}, \ldots)$ with $k<0$ and $a_n \neq \infty$ for all $n>k$,

(c) $(\ldots,a_{l-2},a_{l-1},\infty]$ with $l>0$ and $a_n \neq \infty$ for all $n<l$,

(d) $[\infty,a_{k+1}, \ldots , a_{l-1},\infty]$ with $k<0$, $l>0$ and $a_n \neq \infty$ for all $k<n<l$.

We introduce in S a topology through the neighbourhood basis $\{U_j(a)\}$, $a \in S$, $j \in \mathbb{N}$ where $U_j(a)$ is defined by:

$$U_j(a) = \{ a' \in S : a'_n = a_n \text{ if } |n| < j \} ,$$
$$U_j(a) = \{ a' \in S : a'_n = a_n \text{ if } k < n < j \text{ and } a'_k > j \} ,$$
$$U_j(a) = \{ a' \in S : a'_n = a_n \text{ if } -j < n < l \text{ and } a'_l > j \} ,$$
$$U_j(a) = \{ a' \in S : a'_n = a_n \text{ if } k < n < l \text{ and } a'_k, a'_l > j \} ,$$

according to whether the sequence a is of type (a),(b),(c) or (d), respectively.

Let $\sigma : S \to S$ be the <u>shift automorphism</u> defined by $(\sigma(a))_n = a_{n+1}$. σ is defined on $D(\sigma) = \{ a \in S : a_{-1} \neq \infty \}$. The following lemma is well known (see [DGS]),

<u>LEMMA 18</u>. *With the given topology, S is compact and σ is a homeomorphism with the image.*

There is a bijection between the set of triads given in Corollaries 15 and 16 or 15' and 16', and the set of positive integer numbers $\mathbb{N}$. Let M be an infinity transition matrix with elements $m_{ij} \in \{0,1\}$ for all $(i,j) \in A \times A$, where $m_{ij} = 1$ if and only if the corresponding triads i and j are compatible, otherwise $m_{ij} = 0$.

Let T be the set of all sequences of compatible triads of types (a), (b), (c) and (d). We consider in T the topology induced by S. Let $\bar{\sigma}$ be the restriction of σ on T. Then $(T, \bar{\sigma})$ is a subshift of (S, σ) with transition matrix M (see [DGS]).

<u>LEMMA 19</u>. *T is compact and $\bar{\sigma}$ is a homeomorphism with the image.*

<u>Proof</u>. It is immediate from Lemma 18 and the fact that the complementary of T in S is open.

Q.E.D.

Let D be the subset of the domain of definition of h whose orbits realize a sequence of T. We denote by $\bar{s}(d)$ the sequence realized by the point $d \in D$. Then, from Theorems 17 and 17', it follows easily:

COROLLARY 20. *The map* $\bar{s}:D \longrightarrow T$ *is a continuous surjection and the following diagram commutes*,

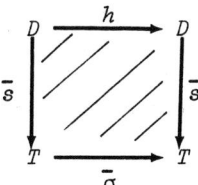

We note that if the map $\bar{s}$ is injective then, by Lemma 19, $\bar{s}$ will be a homeomorphism. In this case, from Corollary 20, $\bar{\sigma}$ is called a subsystem of h. In fact, it seems not easy to prove the injectivity of $\bar{\sigma}$ because this requires a very good knowledge of the global behaviour of the flow.

(IV.8) Gutzwiller's Theorem.

In this section we give a version of Devaney for a theorem of Gutzwiller [G5,6], for more details see [D5] .

Let,

$$S^+ = \{ (q,p) \in I_h : p_2=0,\ q_2>0 \}$$
$$S^- = \{ (q,p) \in I_h : p_2=0,\ q_2<0 \}$$

and $\tilde{S}=S^+\cup S^-$. Let F be the usual Poincaré map on $\tilde{S}$. Of course, F in forward time in not defined on $\tilde{S}\setminus C$ where $C=W^s_-(p^-(0),\mu)\cup W^s_+(p^-(\pi),\mu)\cup W^s_+(p^-(0),\mu)\cup W^s_-(p^-(\pi),\mu)$ and in backward time on $\tilde{S}\setminus E$ where $E=W^u_+(p^+(0),\mu)\cup W^u_-(p^+(\pi),\mu)\cup W^u_-(p^+(0),\mu)\cup W^u_+(P^+(\pi),\mu)$.

Let Λ be the sequences of positive integers of the following four types:

$$(\ldots,s_{-2},s_{-1},s_0,s_1,s_2,\ldots)$$
$$[\infty,s_{-k},\ \cdots\ ,\ s_0,s_1,s_2,\ \ldots)$$
$$(\ldots,s_0,s_1,s_2,\ \cdots\ ,\ s_j,\infty]$$
$$[\infty,s_{-k},\ \cdots\ ,s_0,\ \cdots\ ,\ s_j,\infty]$$

with $j,k\geqslant 1$. We topologize Λ in a similar way to the set S of (IV.7).

Let $p \in \tilde{S}$ and suppose that $F^j(p)$ is defined. We define the jth __passage__
__of__ p to be the segment of orbit containing $F^j(p)$ beginning at the first
prior crossing of $q_2=0$ and ending at the next crossing of the q_1-axis. We
include the endpoints of the orbit segment, even if one or both is the origin. Let $s_j = s_j(p)$ denote the number of times the orbit through p crosses the
q_2-axis during the jth passage. We count collision and ejection as a crossing.

THEOREM 21 (Gutzwiller). _The mapping_ $s : \tilde{S} \longrightarrow \Lambda$ _is a continuous surjection,_
where $(s(p))_j = s_j(p)$ _for_ $p \in \tilde{S}$.

Theorem 21 can be obtained as a Corollary of Theorem 17'. For example,
the orbit which realizes the sequence $[\infty, 1, 5, 1, 1, 1, 2, \infty]$ of Theorem 21 can be obtained from the orbit which realizes the sequence,
$[e(+,u), 1, 5R(-,u)), (R(-,u), 1, R(-,u)), (R(-,u), 2, c(-,u)]$ of Theorem 17'. Note
that instead of $e(+,u)$ we can start with $e(-,u), e(+,l)$ or $e(-,l)$. Thus, in
general, for each sequence of Theorem 21 we have more than one of Theorem
17'.

The map s in Theorem 21 is definitively not 1-1 since the symmetries
of the problem give rise to distinct orbits with the same sequences. In [G6]
Gutzwiller has conjectured that, up to the symmetries, there is a unique solution corresponding to each sequence in Λ. We note that the map $\bar{s}$ of Corollary 20 takes into account these symmetries but it needs the injectivity
in order to obtain $\bar{\sigma}$ as a subsystem of h.

V. THE FLOW ON NEGATIVE ENERGY LEVELS WHEN $1 < \mu \leq 9/8$.

From Proposition II.2 and Corollary II.3, the difference between the cases $1 < \mu \leq 9/8$ and $\mu > 9/8$ is that the equilibrium points $p^{\pm}(\pm\pi/2)$ are sources and sinks without spiraling in the first case and spiral sources and spiral sinks in the second one.

The main result of this chapter will be to prove that the subshift for $\mu > 9/8$, given in Theorem IV.17 and IV.17', disappears totally when $1 < \mu \leq 9/8$. Of course, along this chapter we shall use the notation of Chapter II and IV.

(V.1) The intersection of the invariant manifolds with the surface of section $v=0$.

Lemmas IV.1 and IV.5 are also true for $1 < \mu \leq 9/8$ and the proof of the following proposition is the same as in Theorem IV.3.

PROPOSITION 1. For $1 < \mu \leq 9/8$, if we parametrize the arc $\sigma^u_+(p^+(0),\mu)$ with a parameter $s \in [0,\infty)$ such that $\sigma^u_+(0)=(0,0)$ and $\lim_{s \to \infty} \sigma^u_+(s)=(\pi/2,0)$ then, $\sigma^u_+(s)$ is a continuous arc for all $s \in [0,\infty)$ contained in $\{v'<0\} \cap S$. Furthermore, $\theta(\sigma^u_+(s)) \in [0,\pi)$ for all $s \in [0,\infty)$.

From Figures IV.3 and I.11 respectively, we have that the number of crossings of $\sigma^u_+(p^+(0),\mu)$ with the θ-axis is infinite if $\mu>9/8$ and zero if $\mu=1$.

From now on, if we have a curve Γ and a point p then we define the number of revolutions of Γ around p, $R(\Gamma,p)$, in a similar way to (II.7).

Let $\mu \in (1,9/0]$. The orbit $B^u_+(p^+(0),\mu)$ is forward asymptotic to $p^+(\pi/2)$ (see Theorem II.12). So, from Figures II.4 and II.5 we have that the number of revolutions which it gives around $p^+(\pi/2)$ is zero; see Proposition II.14 and Figure II.14b too. On the other hand, the point $p^+(\pi/2)$ is an hyperbolic point for the flow of the anisotropic Kepler problem given in II.(2); so, from Hartmann's theorem we have that the curve $\Gamma = W_+^u(p^+(0),\mu) \cap S^*$, where S^* is a surface of section transversal to $\gamma_h(\pi/2)$ and close to Λ, is such that $R(\Gamma, \gamma_h(\pi/2) \cap S^*)$ is zero. So, we have that $R(\mu) = R(\sigma_+^u(p^+(0),\mu), (\pi/2,0))$ is finite for all $\mu \in [1,9/8]$. This is what Proposition 2 will say. We shall give another proof using variational equations.

PROPOSITION 2. If $\mu \in (1,9/8]$ then $R(\mu) = R(\sigma_+^u(p^+(0),\mu), (\pi/2,0)) < +\infty$.

Proof. From (II.3) we have that the eigenvalues associated to the equilibrium point $p^+(\pi/2)$ are given by,

$$\lambda_\pm = 2^{-3/2} [-1 \pm (9-8\mu)^{1/2}] \text{ on } \Lambda \text{ and } \lambda = 2^{1/2} \text{ off } \Lambda.$$

On the tangent plane to Λ at the point $p^+(\pi/2)$, (θ, u), the eigenvectors associated to $\lambda_\pm$ are given by, $w_\pm = (-1, -\lambda_\pm)$, see Figures 1. Note that $\lambda_+ = \lambda_-$ if and only if $\mu = 9/8$.

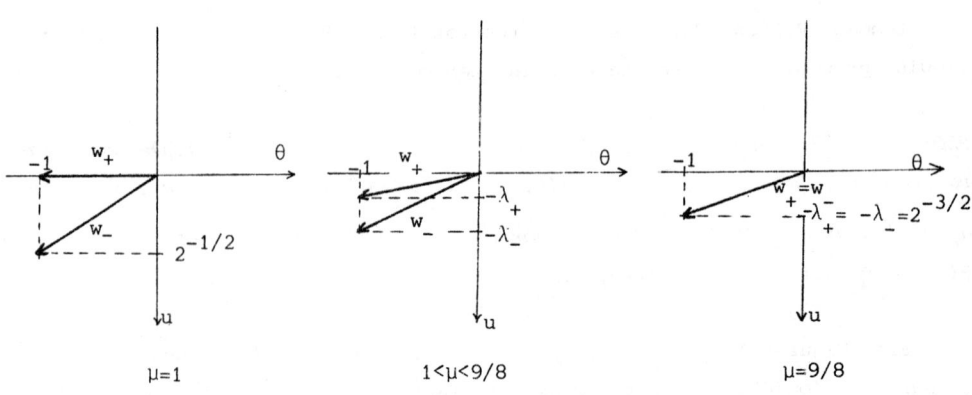

$\mu = 1$ $1 < \mu < 9/8$ $\mu = 9/8$

Figures 1. The eigenvectors $w_\pm = (-1, -\lambda_\pm(\mu))$ for $1 \leq \mu \leq 9/8$.

As we proved in Theorem II.7, the tangent space at a point $p(\tau) \in \gamma_h(\pi/2)$ splits in direct sum of a line L and an ortogonal plane P independently on the point $p(\tau)$. The plane P is generated by the eigenvectors $w_\pm(\tau)$ associated to the eigenvalues $\lambda_\pm(\tau)$ of the matrix,

$$A = \begin{pmatrix} 0 & 1 \\ -V''(\pi/2) & -v(\tau)/2 \end{pmatrix}$$

where $V''(\pi/2) = \mu-1$ and $v(\tau) = -2^{1/2}\tanh(2^{-1/2}\tau)$. In particular, $w_{\pm}(\tau=-\infty) = w_{\pm}$ and $\lambda_{\pm}(\tau=-\infty) = \lambda_{\pm}$ are given in Figures 1.

In P we introduce polar coordinates through $\theta = \rho\cos\Phi$ and $u = \rho\sin\Phi$. Let $\Phi_+(\tau)$ be the angle associated to the eigenvector $w_+(\tau)$ for $\tau \in [-\infty, 0]$.

The orbit $B_+^u(p^+(0),\mu)$ is forward asymptotic to $p^+(\pi/2)$ and if $\tau \to +\infty$ then its tangent vector tends to the strong direction w_+ of Figures 1 with Φ-coordinate equals to $\Phi_+(-\infty)$. So, in order to proof the proposition it is enough to study the function $\Phi_+(\tau)$ when τ increases from $-\infty$ to 0.

In polar coordinates the equation,

$$\eta' = A\eta$$

becomes,

$$\Phi' = (1-\mu)\cos^2\Phi - \sin^2\Phi + 2^{-1/2}\tanh(2^{-1/2}\tau)\sin\Phi\cos\Phi$$

$$\tag{1}$$

$$= (1-\mu) - (2-\mu)\sin^2\Phi + 2^{-1/2}\tanh(2^{-1/2}\tau)\sin\Phi\cos\Phi$$

The initial conditions are $\tau=-\infty$, $\Phi = \Phi_+(-\infty) = \operatorname{atan}(2^{-3/2}[-1+(9-8\mu)^{1/2}]) \in (\pi/2,\pi)$.

If $\tau=-\infty$ then $\Phi' = f(\Phi)+g(\Phi)$ where $f(\Phi) = (1-\mu) - (2-\mu)\sin^2\Phi$ and $g(\Phi) = -2^{-3/2}\sin 2\Phi$, see Figure 2.

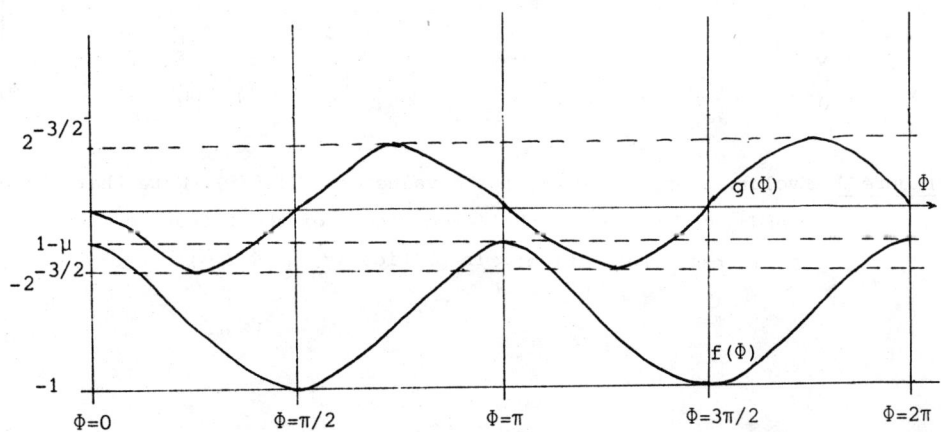

Figure 2. The functions $f(\Phi)$ and $g(\Phi)$.

We consider Φ' restricted to the interval $[0,\pi]$. The same analysis is true in $[\pi,2\pi]$ because Φ' has periodicity equal to π.

If $\mu=9/8$ then is easy to compute that the equation $\Phi'(-\infty) = f(\Phi)+g(\Phi)=0$ has only one zero at $\Phi= \Phi_+(-\infty) = \text{atan}(-2^{-3/2}) \in (\pi/2,\pi)$; furthermore, for $\tau \in (-\infty,0]$, $\Phi'(\tau)<0$. In fact, the function $f(\Phi)$ does not depen on τ and the function $g(\Phi)= g(\Phi,\tau)$ is such that $g(\Phi,\tau)< g(\Phi,-\infty)$ for $\tau\in(-\infty,0]$; so, $\Phi'(\tau)< \Phi'(-\infty)=0$ for $\tau\in(-\infty,0]$.

Suppose $\mu<9/8$. The function $g(\Phi)$ does not depend on μ and $f(\Phi) = f(\Phi,\mu)$ is such that $0>f(\Phi=\pi,\mu)> f(\Phi=\pi,\mu=9/8)$ and $f(\Phi=\pi/2,\mu)= f(\Phi=\pi/2, \mu=9/8)$, see Figure 2. So, if $1<\mu< 9/8$ then the equation $\Phi'(\Phi,\tau=-\infty) = f(\Phi)+g(\Phi)=0$ has two zeros on $(\pi/2,\pi)$ and there exists $\tau_0=\tau_0(\mu)\in (-\infty,0)$ such that $\Phi'(\Phi,\tau)< 0$ for all $\tau\in(\tau_0,0]$ and $\Phi\in(\pi/2,\pi)$. Of course $\Phi'(\Phi,\tau)< 0$ for all $\tau\in(-\infty,0)$ and $\Phi\in [0,\pi/2]$, see Figure 3.

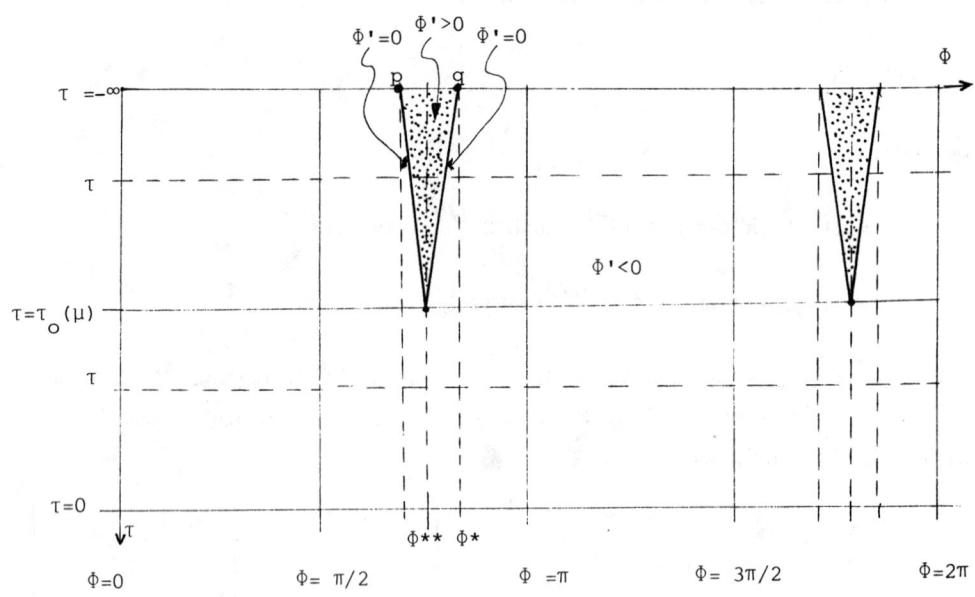

Figure 3. Evolution of $\Phi' = \Phi'(\tau)$ for a value of $\mu\in(1,9/8)$. Note that the points p and q correspond to the Φ-coordinate of the vectors w_- and w_+ given in Figures 1, respectively. In particular, q= $\Phi_+(-\infty)$.

Now, we shall prove the proposition for a fix value of $\mu \in (1, 9/8)$. From Figure 3 we have that the solution $\Phi_+(\tau)$ of (1) is bounded on the interval $[\Phi^{**}, \Phi^*]$ and $\Phi_+' < 0$ for $\tau \in [-\infty\ \tau_o]$. Moreover, from $\tau = \tau_o(\mu)$ to $\tau = 0$ we have finite time and so the function $\Phi_+(\tau)$ decreases a finite value from $\tau = \tau_o$ to $\tau = 0$ too.

When $\mu = 9/8$ we can not use Figure 3 because the shadowed regions do not exist . We shall prove this case in a different way.

If $\mu = 9/8$ then equation (1) becomes,

$$\dot{\Psi} = -9.2^{-5/2} + 7.2^{-5/2} \cos \Psi + \tanh(t) \sin \Psi$$

$$= -9.2^{-5/2} + 7.2^{-5/2} \cos \Psi - \sin \Psi + (1 + \tanh(t)) \sin \Psi \tag{2}$$

where $\Psi = 2\Phi$, $t = 2^{-1/2}\tau$ and $\dot{\Psi} = d\Psi/dt$.

In a neighbourhood of $t = -\infty$, equation (2) can be approximated by,

$$2sz' = Az^2 + Bs \tag{3}$$

where $z = \Psi - 2\ \Phi_+(-\infty)$, A and B are constants, $s = e^{2t}$ and $z' = dz/ds$.
Now, $z(0) = 0$ and the solutions of (3) are bounded by the solutions of the equations,

$$2sz' = (B \pm \varepsilon)s \tag{4}$$

for some $\varepsilon > 0$ in a neighbourhood of $z = 0$ and $s = 0$. Solutions of equations (4) are given by,

$$z = (B \pm \varepsilon)s/2$$

and z increases a finite value when s goes from 0 to a certain value $s_o > 0$. So, Proposition 2 follows for the case $\mu = 9/8$.

Q.E.D.

Numerical computations show that the number of crossings of $\sigma_+^u(p^+(0), \mu)$ with the θ-axis remains equal to zero when μ goes from 1 to 9/8 and that the curve $\sigma_+^u(p^+(0), \mu)$ looks as in Figure 4 for $1 < \mu \leq 9/8$.

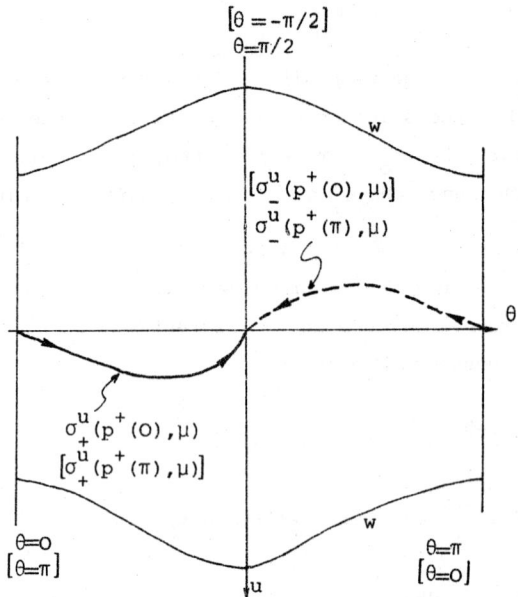

Figure 4. The curve $\sigma_+^u(p^+(0),\mu)$ for $1<\mu\leq9/8$.

As we said before, see Figure 1, the Φ-coordinate of the orbit $B_+^u(p^+(0),\mu)$ when $\tau\to+\infty$ depends on μ in the following way,

$$\Phi(B_+^u(p^+(0),\mu),\tau\to+\infty)=\text{atan}(2^{-3/2}[-1+(9-8\mu)^{1/2}])=\Phi_1(\mu).$$

In Figure 5 it is represented the function (it has been computed numerically),

$$\Phi(\sigma_+^u(p^+(0),\mu),s\to+\infty)=\Phi_2(\mu) \text{ when } \mu\in(1,9/8], \text{ see also Figure 4.}$$

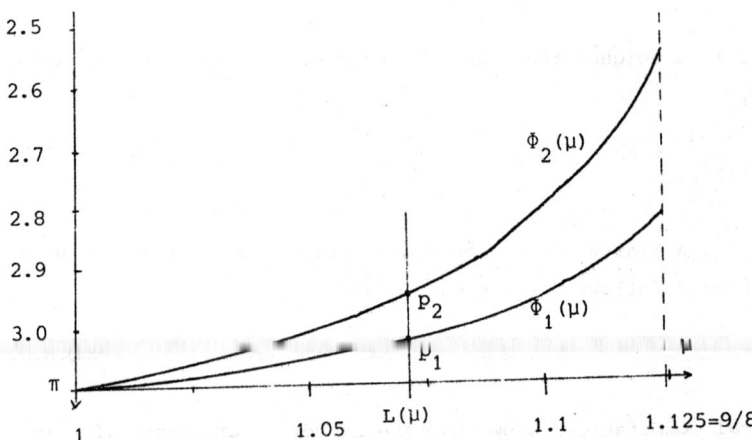

Figure 5. The functions $\Phi_1(\mu)$ and $\Phi_2(\mu)$ for $\mu\in[1,9/8]$. Here, for $p_1,p_2\in L(\mu)$, p_2-p_1 is the angle rotated when we follow $W_+^u(p^+(0),\mu)$ from Λ to $\{v=0\}$.

(V.2) Dynamical description.

Let $\mu \in [1,9/8]$ and let γ be an orbit of the anisotropic Kepler problem. From Proposition 2 and (IV.5) the number of crossings of γ with the heavy axis (q_2) between two consecutive crossings with the q_1-axis is bounded by some $n^* \in \mathbb{N}$. Figure 4 shows that $n^*=2$ for all $\mu \in [1,9/8)$.

VI. SYMMETRIC PERIODIC ORBITS

(VI.1) Definitions and preliminary results.

From (II.2) the anisotropic Kepler problem has the symmetries,

$$S_o : (q_1,q_2,p_1,p_2,t) \longrightarrow (q_1,q_2,-p_1,-p_2,-t)$$
$$S_1 : (q_1,q_2,p_1,p_2,t) \longrightarrow (q_1,-q_2,-p_1,p_2,-t)$$
$$S_2 : (q_1,q_2,p_1,p_2,t) \longrightarrow (-q_1,q_2,p_1,-p_2,-t)$$
$$S_3 = S_2 \circ S_1 : (q_1,q_2,p_1,p_2,t) \longrightarrow (-q_1,-q_2,-p_1,-p_2,t)$$
$$S_4 = S_2 \circ S_o : (q_1,q_2,p_1,p_2,t) \longrightarrow (-q_1,q_2,-p_1,p_2,t)$$
$$S_5 = S_1 \circ S_o : (q_1,q_2,p_1,p_2,t) \longrightarrow (q_1,-q_2,p_1,-p_2,t)$$

They can be interpreted in the following way.

Let $\gamma(t) = (q_1(t),q_2(t),p_1(t),p_2(t))$ be a solution of II.(1). Then, $S_o(\gamma(t)) = (q_1(-t),q_2(-t),-p_1(-t),-p_2(-t))$ is another solution. In figure 1 we draw all the solutions $S_i(\gamma(t))$ for $i=0,1,2,3,4,5$.

For $i \in \{0,1,2,3,4,5\}$ the orbit $\gamma(t)$ will be called S_i-symmetric if and only if $S_i(\gamma(t)) = \gamma(t)$.

LEMMA 1. _(i) For $i=1,2$ we have that an orbit $\gamma(t)$ is S_i-symmetric if and only if it crosses the q_i-axis ortogonally._
(ii) An orbit $\gamma(t)$ is S_o-symmetric if and only if it has a point on the zero velocity curve.
(iii) For $i=4,5$, an orbit $\gamma(t)$ is S_i-symmetric if and only if it is S_o-symmetric
(iv) All the S_3-symmetric orbits are periodic.

The proof of this lemma follows easily. Note that (i),(ii) and(iii) characterize the S_i-symmetric orbits for $i=0,1,2,4,5$.

We are interested in the S_i-symmetric periodic orbits. These types of orbits were studied by Birkhoff [B] and De Vogelaere [De] for other Hamiltonian systems. More recently Devaney in [D1] proved (i) of the following proposition, but (ii) follows in a similar way.

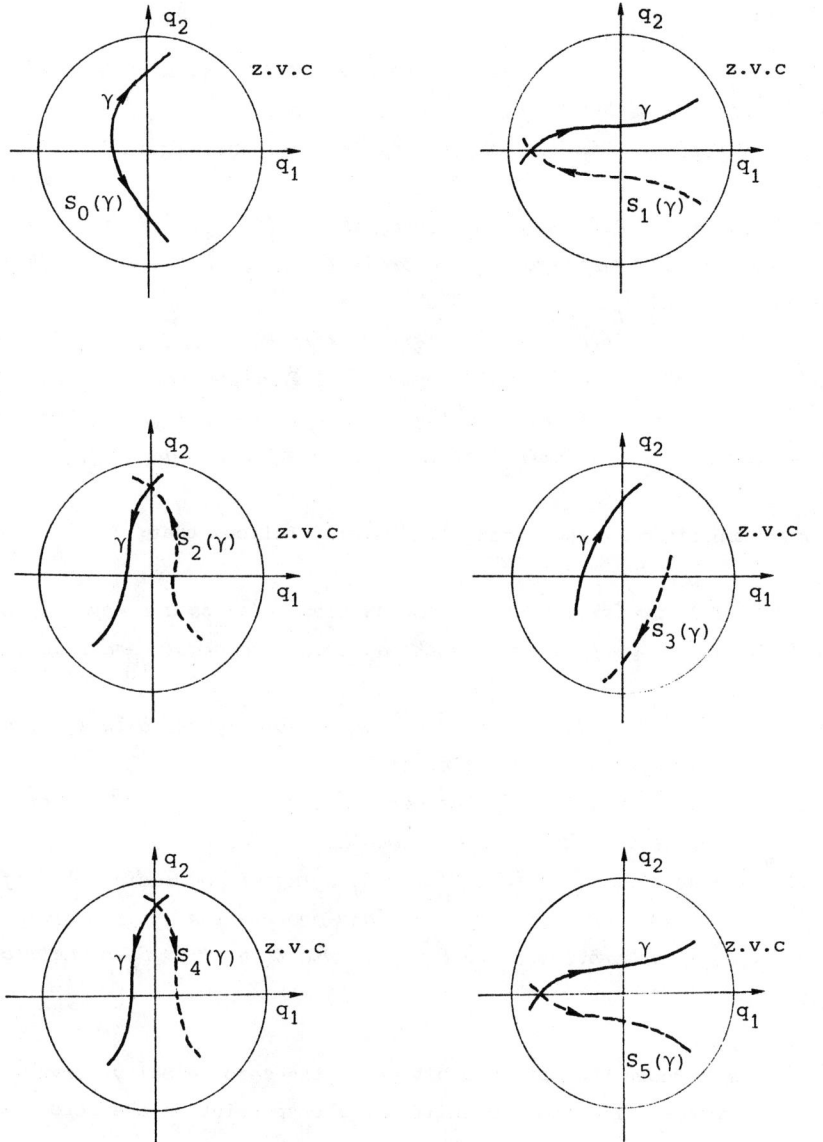

<u>Figure 1</u>. The symmetric orbits of $\gamma(t)$: $S_i(\gamma(t))$ for i=0,1,2,3,4,5.

PROPOSITION 2. Let X be a R_1 and R_2-reversible vector field on a $2n$-dimensional manifold M. We denote by φ_t the flow of X. Suppose that x and $\varphi_T(x) \in Fix(R_1) \cup Fix(R_2)$, where T is the smallest value of $t>0$ with this property.

(i) For $i=1,2$, if x, $\varphi_T(x) \in Fix(R_i)$ then, the orbit $\varphi_t(x)$ is periodic of period $2T$ and R_i-symmetric. Furthermore, the orbit $\varphi_t(x)$ meets the set $Fix(R_i)$ exactly in the points x and $\varphi_T(x)$.

(ii) If $(x, \varphi_T(x)) \in Fix(R_1) \times Fix(R_2) \cup Fix(R_2) \times Fix(R_1)$ then, the orbit $\varphi_t(x)$ is periodic of period $4T$ and R_1 and R_2 symmetric. Furthermore, the orbit $\varphi_t(x)$ meets the set $Fix(R_1) \cup Fix(R_2)$ exactly at the points x, $\varphi_T(x)$, $\varphi_{2T}(x)$ and $\varphi_{3T}(x)$, where x, $\varphi_{2T}(x) \in Fix(R_i)$, $\varphi_T(x)$, $\varphi_{3T}(x) \in Fix(R_j)$ and $i \neq j$.

From Proposition 2 and Lemma 1 it clearly follows that:

COROLLARY 3. (i) For $i=1,2$ we have that an orbit $\gamma(t)$ is a S_i-symmetric periodic orbit if and only if it crosses the q_i-axis ortogonally and exactly at two points.

(ii) An orbit $\gamma(t)$ is a S_o-symmetric periodic orbit if and only if it meets the zero velocity curve exactly at two points.

(iii) An orbit $\gamma(t)$ is a S_1 and S_2-symmetric periodic orbit if and only if it crosses the q_1-axis and the q_2-axis ortogonally.

(iv) For $i=1,2$ an orbit $\gamma(t)$ is a S_o and S_i-symmetric periodic orbit if and only if it meets the zero velocity curve and crosses the q_i-axis ortogonally.

(v) For $i=4,5$, if an orbit $\gamma(t)$ is S_i-symmetric then it is S_o-symmetric and periodic.

It is well known that if an orbit meets the zero velocity curve it has to be in the normal direction. Then it has a cusp point on the zero velocity curve.

Proposition 2 and Corollary 3 give us a technique in order to obtain symmetric periodic orbits (s.p.o) with respect to S_o, S_1 or S_2. We shall prove that there are s.p.o with respect to S_1 and S_2, so with respect to S_3.

(VI.2) The case μ=1.

When μ=1 we have the Kepler problem. From Chapter I and (VI.1) it follows that:

PROPOSITION 4. (i) There is a bijection between the symmetric orbits (but not periodic) with respect to S_o and the circle. They are elliptic ejection-collision orbits. See Figure 2.

(ii) For i=1,2 there is a bijection between the s.p.o with respect to S_i and two copies of the segment (0,-1/h). One copy corresponds to the direct ellipses and the other one to the retrograde ellipses. See Figure 3.

(iii) There is a bijection between the s.p.o with respect to S_1 and S_2 and the two points $\pm$ $(2h)^{-1}$. They correspond to circular orbits. See Figure 4.

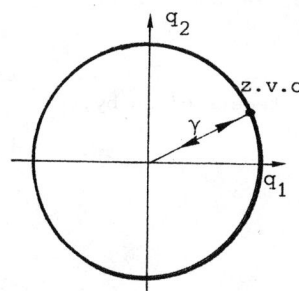

Figure 2. S_o-symmetric orbits

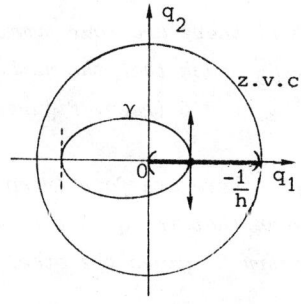
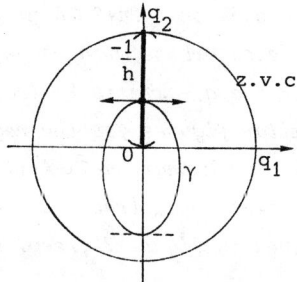

Figure 3. S_i-symmetric orbits for i=1,2.

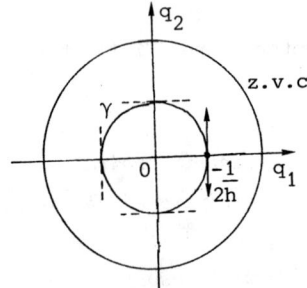

<u>Figure 4</u>. Symmetric orbits with respect to S_1, S_2 and S_3.

Note that the case $\mu=1$ is a degenerated case because a continuous of symmetries is possible

<u>(VI.3) The case $\mu>9/8$.</u>

We denote by $A(\mu)$ the subset of the positive integers given by,

$$A(\mu) = \{1, 2, \ldots, \ldots\} \text{ if } 9/8 < \mu < \mu_c \quad \text{and}$$
$$A(\mu) = \{2n_o-1, 2n_o, \ldots, \cdot\} \text{ if } \mu \geqslant \mu_c$$

where $n_o = n_o(\mu)$ is defined in Theorem IV.13.

<u>THEOREM 5</u>. *If $\mu>9/8$ then the following holds.*
(i) For each n such that $2n+2 \in A(\mu)$ (resp. $2n+1 \in A(\mu)$) there are four symmetric ejection-collision orbits with respect to S_o (resp. S_2) such that the number of crossings with the q_2-axis is $2n$ (resp. $2n-1$). See Figures 5 (resp. Figures 6). There are similar figures for the region $q_2 \leqq 0$.
(ii) For each n such that $2n-1 \in A(\mu)$ (resp. $2n \in A(\mu)$) there are four (resp. two) orbits with respect to S_2 (resp. S_o and S_2, or S_4) such that the qualitative be-haviour is given in Figure 7 (resp. Figure 8). Symmetry S_3 gives the other two orbits.
(iii) For each n and m such that $2n-1, 2m-1 \in A(\mu)$ (resp. $2n, 2m \in A(\mu)$) there are two s.p.o with respect to S_2 (resp. S_o) such that the qualitative behaviour is shown in Figure 9 (resp Figure 10). When n=m the orbit is also symmetric with respect to S_1; so, S_3-symmetric (resp. S_5-symmetric).

(iv) For each n and m such that 2n-1, 2m ∈ A(μ) there are two s.p.o with respect to S_0 and S_2 such that the qualitative behaviour is shown in Figure 11. Symmetry S_3 gives the other orbit.

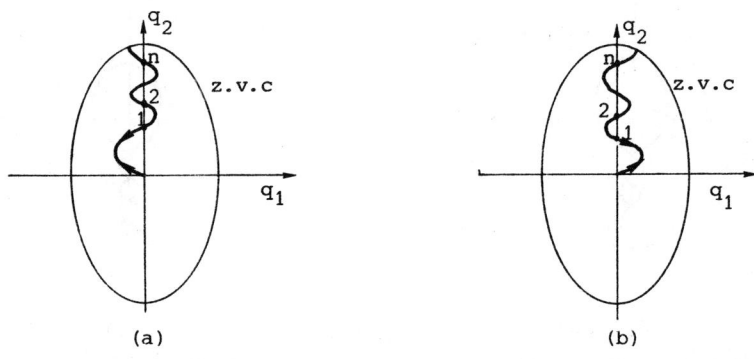

(a)   (b)

Figures 5. (a): S_0-symmetric orbit. This orbit realizes the sequence $\{T_n\}$
where $T_0 = [e(+,u),2n+2,c(-,u)]$. (b): S_0-symmetric orbit. This
orbit realizes the sequence $\{T_n\}$ where $T_0 = [e(-,u),2n+2,c(+,u)]$.

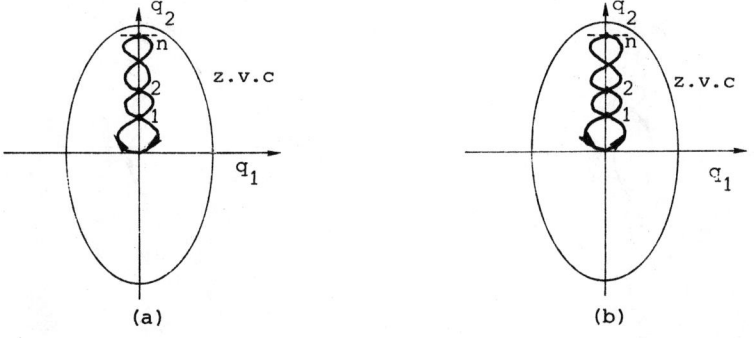

(a)   (b)

Figures 6. (a): S_2-symmetric orbit. This orbit realizes the sequence $\{T_n\}$
where $T_0 = [e(+,u),2n+1,c(+,u)]$. (b): S_2-symmetric orbit. This
orbit realizes the sequence $\{T_n\}$ where $T_0 = [e(-,u),2n+1,c(-,u)]$.

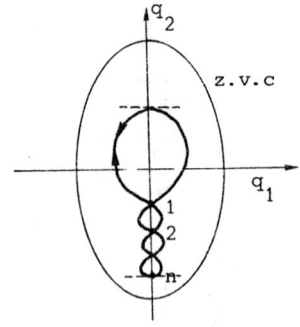

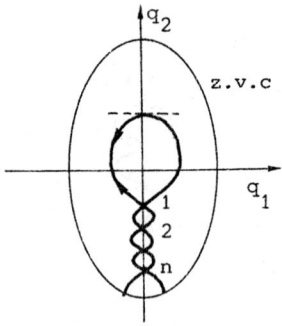

Figure 7. The two S_2-symmetric p.o
These p.o realize the se-
quence $\{T_k\}$ where T_k=
$(R(+,1),2n-1,R(+,1))$ or
$(R(-,1),2n-1,R(-,1))$ for
all $k \in N$.

Figure 8. S_o and S_2 symmetric p.o.
This p.o. realizes the se-
quence $\{T_k\}$ where $T_k = T_{k+2}$ and
$T_k T_{k+1}$=
$(R(+,1),2n,R(-,1))(R(-,1),2n,R(+,1))$
or $(R(-,1),2n,R(+,1))(R(+,1),2n,R(-,1))$,
for all $k \in N$.

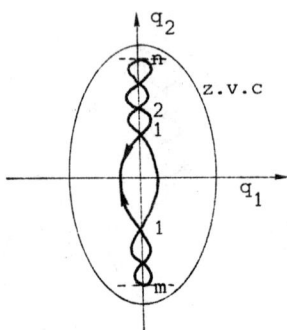

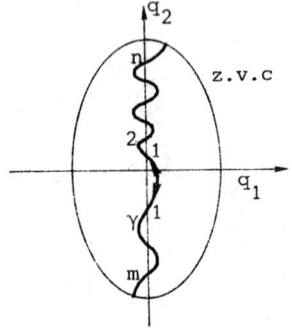

Figure 9. The two S_2-symmetric p.o
These p.o realize the se-
quence $\{T_k\}$ where $T_k = T_{k+2}$
and $T_k T_{k+1}$ =
$(C(-,u),2m-1,C(-,1))(C(-,1),2n-1,C(-,u))$ or
$(C(+,u),2m-1,C(+,1))(C(+,1),2n-1,C(+,u))$
for all $k \in N$.

Figure 10. S_o-symmetric p.o. This p.o.
realizes the sequence $\{T_k\}$
where $T_k = T_{k+2}$ and $T_k T_{k+1}$ =
$(C(-,u),2m,C(-,1))(C(-,1),2n,C(-,u))$.
The other S_o-symmetric p.o. is
$S_2(\gamma)$.

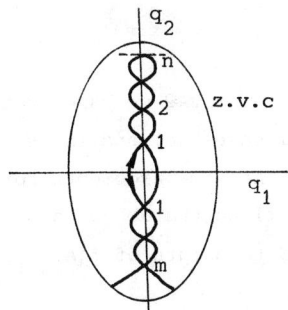

Figure 11. S_o and S_2 symmetric p.o. This p.o realizes the sequence $\{T_k\}$ where
$T_k = T_{k+4}$ and $T_k T_{k+1} T_{k+2} T_{k+3} =$
$(C(-,u),2m,C(+,1))(C(+,1),2n-1,C(+,u))(C(+,u),2m,C(-,1))(C(-,1),2n-1,C(-,u))$.

Proof. We consider $\mu > 9/8$.

(i): The existence of s.p.o with respect to S_o follows from the fact that $\sigma_+^u(p^+(0),\mu)$ (resp. $\sigma_+^u(p^+(\pi),\mu)$) cuts $\sigma_-^s(p^-(0),\mu)$ (resp. $\sigma_-^s(p^-(\pi),\mu)$) and $\sigma_-^u(p^+(\pi),\mu)$ (resp. $\sigma_-^u(p^+(0),\mu)$) cuts $\sigma_+^s(p^-(\pi),\mu)$ (resp. $\sigma_+^s(p^-(0),\mu)$) on $\{u=0\} \cap S$, see Figure IV.22. The existence of s.p.o with respect to S_2 follows from the fact that $\sigma_+^u(p^+(0),\mu)$ (resp. $\sigma_+^u(p^+(\pi),\mu)$) cuts $\sigma_+^s(p^-(\pi),\mu)$ (resp. $\sigma_+^s(p^-(0),\mu)$) and $\sigma_-^u(p^+(\pi),\mu)$ (resp. $\sigma_-^u(p^+(0),\mu)$) cuts $\sigma_-^s(p^-(0),\mu)$ (resp. $\sigma_-^s(p^-(\pi),\mu)$) on $\{\theta = \pm \pi/2\} \cap S$, see Figure IV.22.

(ii) and (iii): We define the families of segments $\{a_r\}$, $\{b_r\}$, $\{c_r\}$ and $\{d_r\}$ for $r \geq 2$ like in Figure 12.

(a) In a similar way to the proof of Lemma IV.7 we have that $g(a_r)$ for $r \geq 2$ and r even and $g(b_r)$ for $r \geq 2$ and r odd meets $\{\theta = \pi, u < 0\} \cap S$ and $\{\theta = -\pi/2, u < 0\} \cap S$, see Figure 13. By Corollary 13 we obtain:

(a.1) If $r \geq 2$ is odd then $g(b_r) \cap \{\theta = \pi\} \cap S$ are s.p.o with respect to S_1 and S_2, and $g(b_r) \cap \{\theta = -\pi/2\} \cap S$ are s.p.o with respect to S_2.

(a.2) If $r \geq 2$ is even then $g(a_r) \cap \{\theta = \pi\} \cap S$ are s.p.o with respect to S_o and S_1, and $g(a_r) \cap \{\theta = -\pi/2\} \cap S$ are s.p.o with respect to S_o and S_2.

(b) As above, $g(a_r)$ for $r \geq 2$ and r odd and $g(b_r)$ for $r \geq 2$ and r even, meets $\{\theta = 0, u > 0\} \cap S$ and $\{\theta = -\pi/2, u > 0\} \cap S$ like in Figure 14. Again, by Corollary 13 we have:

(b.1) If $r \geq 2$ is odd then $g(a_r) \cap \{\theta = 0\} \cap S$ are s.p.o with respect to S_1 and S_2, and $g(a_r) \cap \{\theta = -\pi/2\} \cap S$ are s.p.o with respect to S_2.

(b.2) If $r \geq 2$ is even then $g(b_r) \cap \{\theta = 0\} \cap S$ are s.p.o with respect to S_o and S_1, and $g(b_r) \cap \{\theta = -\pi/2\} \cap S$ are s.p.o with respect to S_o and S_2.

(c) By using the symmetry S_3, for $r \geqslant 2$ we have that the s.p.o obtained from $g(c_r)$ and $g(d_r)$ follows from cases (a) and (b) respectively.

We note that if we use the same arguments for $f(a_r)$, $f(b_r)$, $f(c_r)$ and $f(d_r)$, the s.p.o obtained will be the above ones.

(iv): Here, s.p.o correspond to points of $f(A_{2n-1}) \cap g(B_{2n})$ where $A, B \in \{a, b, c, d\}$.

Q.E.D.

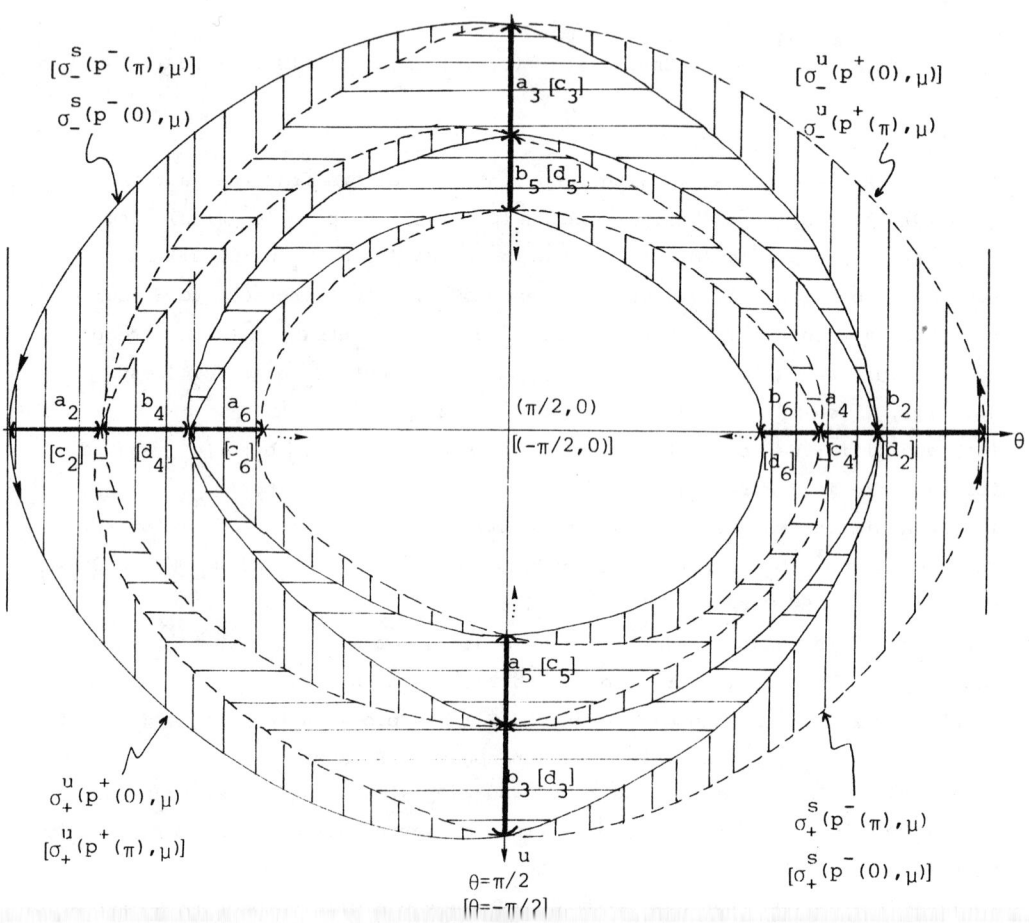

Figure 12. The families of segments $\{a_r\}_{r \in \mathbb{N}}$, $\{b_r\}_{r \in \mathbb{N}}$, $\{c_r\}_{r \in \mathbb{N}}$ and $\{d_r\}_{r \in \mathbb{N}}$ (compare with Figures IV.22).

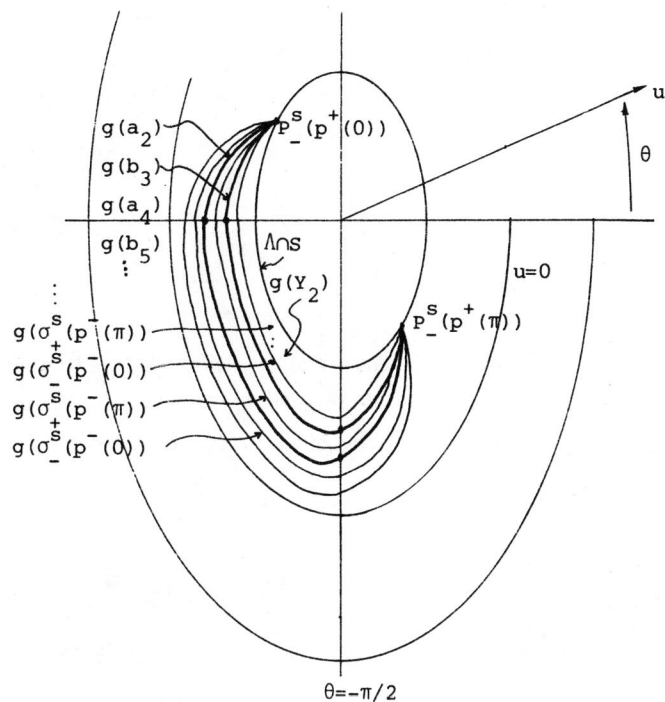

Figure 13 . The families of curves $g(a_r)$ for r even and $g(b_r)$ for r odd,
(compare with Figure IV.12 and IV.14).

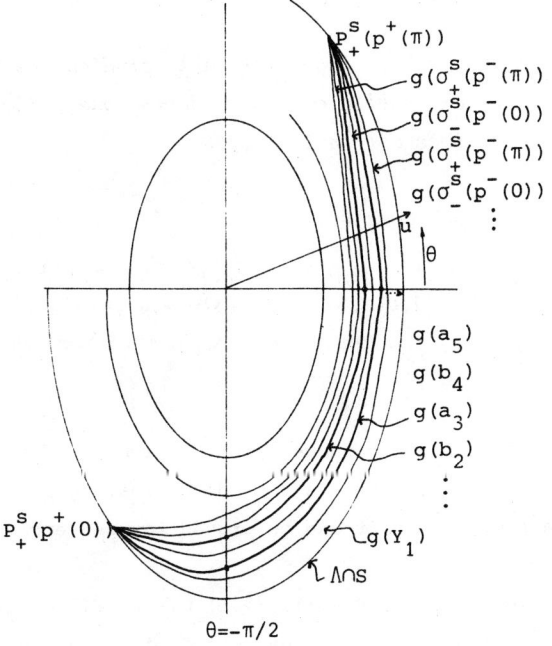

Figure 14 . The families of curves $g(a_r)$ for r odd and $g(b_r)$ for r even
(compare with Figure IV.12 and IV.14).

REMARK 1. *Theorem 5 classifies the qualitative behaviour of s.p.o obtained from the points of,*

$$f(A_r) \cap C,$$
$$f(B_r) \cap C \quad and \tag{1}$$
$$f(A_{r_1}) \cap g(B_{r_2})$$

where $A, B \in \{a, b, c, d\}$, a, b, c and d are the segments defined in the proof of Theorem 5 and C is one of the semi-axes $\{\theta=0\} \cap S$, $\{\theta = \pi\} \cap S$ or $\{\theta = \pm \pi/2\} \cap S$. In a similar way, we could classify the qualitative behaviour of s.p.o corresponding to points of,

$$h^s(f(A_r) \cap C),$$
$$h^{-s}(g(A_r) \cap C) \quad and \tag{2}$$
$$h^s(f(A_{r_1})) \cap h^{-t}(g(B_{r_2}))$$

for $s, t \geq 1$. The existence of s.p.o of type (2) will follow from Theorem 6.

REMARK 2. *Symmetric periodic orbits described in Figure 10 when either n or m equals 1 were obtained by Devaney in [D5].*

REMARK 3. *Gutzwiller in [G7] has numerically studied the periodic orbits in the anisotropic Kepler problem such that during one period they cross the heavy axis, q_2, $2n$ times with $n=1, 2, 3, 4, 5$.*

THEOREM 6. *For each periodic sequence $\{T_n\}$ given in Theorems IV.17 and IV.17' such that if $T_{-s}T_{-s+1}, \ldots, T_o, \ldots, T_{t-1}T_t$ is its period and $T_t=(A, m, B)$ then $A, B \in \{C(+, u), C(-, u), C(+, l), C(-, l)\}$ or $A, B \in \{R(+, u), R(-, u), R(+, l), R(-, l)\}$, there exists a s.p.o of the anisotropic Kepler problem which realizes it.*

Proof. Let $\{T_n\}$ be a periodic sequence of Theorem IV.17 and let $T_{-s}T_{-s+1}, \ldots, T_o, \ldots, T_{t-1}T_t$ be its period. We shall prove Theorem 6 for this sequence; in a similar way it can be proved for sequences of Theorem IV.17'.

Let $A, B \in \{R(-, u), R(+, u), R(-, l), R(+, l), C(-, u), C(+, u), C(-, l), C(+, l)\}$ and let $(g^{-1}(A) \cap f^{-1}(B), n)$ be one of the sets defined in the proof of Theorem IV.14 and shown in Figure IV.35.

We define the set $S(A,n,B)=(g^{-1}(A) \cap f^{-1}(B),n) \cap (\{u=0\} \cup \{\theta=\pi/2\} \cup \{\theta=-\pi/2\})$.
Note that, by definition, $S(A,n,B) = \emptyset$ if and only if,

$(A,B) \in \{(C(+,1),R(+,u)),(C(+,u),R(+,1)),(R(-,u),C(+,u)),(R(-,1),C(+,1)),$
$(C(-,1);R(-,u)),(C(-,u),R(-,1)),(R(+,u),C(+,u)),(R(+,1),C(+,1)),C(+,1),R(-,u)),$
$(C(+,u),R(-,1)),(R(-,u),C(+,u)),(R(-,1),C(+,1)),(C(-,1),R(+,u)),C(-,u),R(+,1)),$
$(R(+,u),C(-,u)),(R(+,1),C(-,1))\}=X.$ So, if (A,n,B) is such that $(A,B) \in X$ then,
$T_t \neq (A,n,B)$.

By using the same arguments as Theorem IV.17, the set :

$$Z = T_{-s} \cap h^{-1}(T_{-s+1}) \cap \ldots \cap h^{-s}(T_0) \cap \ldots \cap h^{-s-t+1}(T_{t-1}) \cap h^{-s-t}(S')$$

where $S' = g^{-1}(S(T_t))$, is an arc going from the point $P^s(\ ,\mu)$ to the opposite
side in the triangular sector C if $T_{-s} = (C,n',D)$. So, there exists at least
one point $p \in Z \cap (\{\theta=-\pi/2\} \cup \{\theta=0\} \cup \{\theta=\pi/2\} \cup \{\theta=\pi\})$, see Figures IV.31 and
IV.32.

By Corollary 3, the orbit through p is a s.p.o and then realizes the
sequence $\{T_n\}$.

Q.E.D.

APPENDIX

The evolution of the arc $\sigma_+^u(\mu) = \sigma_+^u(p^+(0),\mu)$ is partially showed in Figures 7 of Chapter IV. It has also been computed numerically. So, we can change Theorem IV.10 by:

$$S_{(\pm\pi/2,\mu)}(U(\pm\pi/2,\mu)) = \{2,3,4, \ldots\} \qquad \text{if } \mu \geq \mu_c.$$

Also, in Corollary IV.15 we have to add the sets corresponding to $A_{8j+i} = (X,k)$ for $i \in \{1,2,3,4,5,6,7,8\}$, $j=1,2,3,\ldots$ and $2 \leq k \leq 2n_1$.
In Corollary IV.16 we have the triads of type $[A,2n,B]$ for $n=1, \ldots, n_1$.

REFERENCES

[A1] V.M. Alekseev, Quasirandom dynamical systems I,II,II, Math.USSR-Sbornic 5, pp.73-128(1968);6, pp.505-560(1968);7, pp.1-43(1969).

[A2] V.M. Alekseev, Symbolic dynamics, 11th Math. School, ed.Mitropolskii and Samoilenko, Kiev 1976 (Russian).

[AM] R. Abraham and J.Marsden, Foundations of Mechanics, Reading Mass.: Benjamin/Cummings, 1978.

[B] G. Birkhoff, Dynamical Systems with two degrees of freedom, Trans.Am.Math. Soc., Vol.18(1917), pp. 199-300.

[CLL] J. Casasayas and J. Llibre, Invariant manifolds associated to homothetic orbits in the n-body problem, Indiana University Math. Jour. 31(1982), pp. 463-470.

[CPR] R.C. Churchill, G. Pecelli and D.L. Rod, Isolate unstable periodic orbits, Jour. Differential Equations, 17(1975), pp. 329-348.

[D1] R.L. Devaney, Reversible diffeomorphisms and flows, Transactions Am. Math. Soc., Vol. 218(1976), pp. 90-113.

[D2] R.L. Devaney, Collision Orbits in the Anisotropic Kepler Problem, Invent. Math., No.45(1978), pp. 221-251.

[D3] R.L. Devaney, Nonregularizability of the Anisotropic Kepler Problem, Jour. Differential Equations, 29(1978), pp. 253-268.

[D4] R.L. Devaney, Transverse Heteroclinic orbits in the Anisotropic Kepler Problem, Lecture Notes in Math., No.668, Springer-Verlag, 1978, pp. 67-87.

[D5] R.L. Devaney, Singularities in classical Mechanical Systems, Progress in Math., Vol. 10, Birkhäser, 1981, pp. 211-333.

[De] R. DeVogelaere, On the Structure of symmetric periodic solutions of conservative systems, with applications, Contributions to the Theory on Nonlinear Oscilations, Vol. IV, Ann. of Math. Studies No. 41, Princeton Univ. Press, Princeton N.J., 1958, pp. 53-84.

[DGS] M. Denker, C. Grillembergh and K. Sigmund, Ergodic Theory on Compact
      Spaces, Lecture Notes in Math., No. 527, Springer-Verlag, 1976.

[E]   R. Easton, Regularization of vector fields by suggery, Jour. Differen-
      tial Equations 10(1971), pp. 92-99.

[G1] M.C. Gutzwiller, Phase-Integral Approximation in Momentum Space and
      the Bound States of an Atom, Jour. of Math. Physics, Vol.8, No. 10
      (1967), pp. 1979-2000.

[G2] MC. Gutzwiller, Phase Integral Approximation in Momentum Space and
      the Bound States of an Atom II, Jour. of Math. Physics, Vol. 10, No. 6
      (1969), pp. 1004-1020.

[G3] M.C. Gutzwiller, Energy Spectrum According to Classical Mechanics,
      Jour. of Math. Physics, Vol. 11, No. 6(1970), pp. 1971-1806.

[G4] M.C. Gutzwiller, Periodic orbits and Classical Quantization Condition,
      Jour. of Math. Physics, Vol. 12, No. 3(1971), pp. 343-358.

[G5] M.C. Gutzwiller, The Anisotropic Kepler Problem in two dimensions, Jour.
      of Math. Physics, Vol. 14 (1973), pp.139-152.

[G6] M.C. Gutzwiller, Bernouilli sequences and trajectories in the Anisotro-
      pic Kepler Problem, Jour. of Math. Physics, Vol. 18(1977), pp. 806-823.

[G7] M.C. Gutzwiller, Periodic orbits in the Anisotropic Kepler Problem,
      Classical Mechanics and Dynamical Systems, Marcel Dekker, 1982, pp. 69-89.

[HPS] M. Hirsch, C. Pugh and M. Shub, Invariant manifolds, Lectures Notes in
      Math., No. 583, Springer-Verlag, 1977.

[LS] E. Lacomba and C. Simó, Bounding manifolds for energy surfaces in Celes-
      tial Mechanics Problems, Celestial Mechanics 28(1982), pp. 37-48.

[LLS] J. Llibre and C. Simó, Characterization of Transversal homothetic solu-
      tions in the n-body problem, Archive for Rational Mechanics and Analy-
      sis, Vol. 77, No. 2 (1981).

[LMS] J. Llibre, R. Martinez and C. Simó, Qualitative study of the planar
      isosceles three-body problem, prepint, 1983.

[Mc] R. Mc.Gehee, Triple collision in the collinear three-body Problem, Inv.
      Math. 27(1974), pp. 191-227.

[Mo] J. Moser, Stable and random motions in dynamical systems, Princeton
      Univ. Press (Study 77), N.J. University Press, 1973.

[S1]   S. Smale, Differentiable dynamical systems, Bull. Amer. Math. Soc. 73
       (1967), pp. 747-817.

[S2]   S. Smale, Topology and Mechanics I, Invent, Math. 10, pp.305-331.

General instructions to authors for
PREPARING REPRODUCTION COPY FOR MEMOIRS

For more detailed instructions send for AMS booklet, "A Guide for Authors of Memoirs."
Write to Editorial Offices, American Mathematical Society, P. O. Box 6248,
Providence, R. I. 02940.

MEMOIRS are printed by photo-offset from camera copy fully prepared by the author. This means that, except for a reduction in size of 20 to 30%, the finished book will look exactly like the copy submitted. Thus the author will want to use a good quality typewriter with a new, medium-inked black ribbon, and submit clean copy on the appropriate model paper.

Model Paper, provided at no cost by the AMS, is paper marked with blue lines that confine the copy to the appropriate size. Author should specify, when ordering, whether typewriter to be used has PICA-size (10 characters to the inch) or ELITE-size type (12 characters to the inch).

Line Spacing — For best appearance, and economy, a typewriter equipped with a half-space ratchet — 12 notches to the inch — should be used. (This may be purchased and attached at small cost.) Three notches make the desired spacing, which is equivalent to 1-1/2 ordinary single spaces. Where copy has a great many subscripts and superscripts, however, double spacing should be used.

Special Characters may be filled in carefully freehand, using dense black ink, or INSTANT ("rub-on") LETTERING may be used. AMS has a sheet of several hundred most-used symbols and letters which may be purchased for $5.

Diagrams may be drawn in black ink either directly on the model sheet, or on a separate sheet and pasted with rubber cement into spaces left for them in the text. Ballpoint pen is *not* acceptable.

Page Headings (Running Heads) should be centered, in CAPITAL LETTERS (preferably), at the top of the page — just above the blue line and touching it.

LEFT-hand, EVEN-numbered pages should be headed with the AUTHOR'S NAME;
RIGHT-hand, ODD-numbered pages should be headed with the TITLE of the paper (in shortened form if necessary).
Exceptions: PAGE 1 and any other page that carries a display title require NO RUNNING HEADS.

Page Numbers should be at the top of the page, on the same line with the running heads.

LEFT-hand, EVEN numbers — flush with left margin;
RIGHT-hand, ODD numbers — flush with right margin.
Exceptions: PAGE 1 and any other page that carries a display title should have page number, centered below the text, on blue line provided.

FRONT MATTER PAGES should be numbered with Roman numerals (lower case), positioned below text in same manner as described above.

MEMOIRS FORMAT

It is suggested that the material be arranged in pages as indicated below.
Note: Starred items (*) are requirements of publication.

Front Matter (first pages in book, preceding main body of text).

Page i — *Title, *Author's name.

Page iii — Table of contents.

Page iv — *Abstract (at least 1 sentence and at most 300 words).

*1980 Mathematics Subject Classifications represent the primary and secondary subjects of the paper. For the classification scheme, see Annual Subject Indexes of MATHEMATICAL REVIEWS beginning in December 1978.

Key words and phrases, if desired. (A list which covers the content of the paper adequately enough to be useful for an information retrieval system.)

Page v, etc. — Preface, introduction, or any other matter not belonging in body of text.

Page 1 — Chapter Title (dropped 1 inch from top line, and centered).

Beginning of Text.

Footnotes: *Received by the editor date.
Support information — grants, credits, etc.

Last Page (at bottom) — Author's affiliation.

ABCDEFGHIJ—AMS—8987654